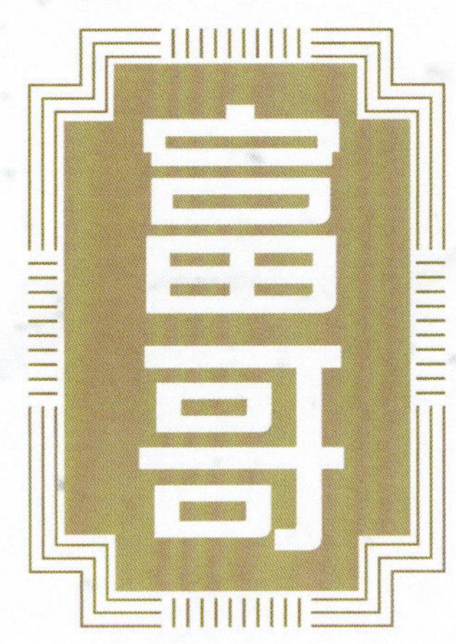

富哥𠮶人坊

細嘗粵菜真味

鄭錦富・方曉嵐 著

萬里機構

推薦序一

無隅之鮮，藏於富哥一匙一味之間

世間的美味千萬種，有些驚艷，有些懷念，而極少數，能讓人第一次已扣住心弦。我走遍世界，品嘗過近千顆米芝蓮星光，卻始終忘不了香港一位隱世廚神——鄭錦富（富哥）。

十多年前，我初訪其餐廳，當時高朋滿座、久候無位，正準備離去時，恰巧在電梯裏與富哥相遇。他得知情況後，親自安排座位、引我入席，溫厚無架子，讓我想起好友巴黎三星主廚 Pierre Gagnaire 聖誕夜的款待，那種不着痕跡的體貼，甚至被邀請到廚房雅座……一如富哥總令人銘記。

此後，每趟香港之行，我總把最後一餐留給富哥。他的廚藝沉着穩健，刀工如書法般簡潔利落，而最打動人心的，是他對食材所懷抱的敬意。從不誇張炫技，而是以極簡之道，喚醒食材最純粹的鮮美。有次我與友人相談稍久，富哥將已上桌的菜式收回重出，只因「冷了，味道就不對了」。這份堅持，令人尊敬。

他不僅是名廚，更是無私的師者。有回我在香港嚐到上好的牛腩清湯，與他分享後，富哥竟親自熬湯款待，毫不保留。這樣的情誼，溫潤如湯，在我心裏久久不散。

富哥的料理哲學，與我所倡「無隅」藝術精神不謀而合——無界、無我、無限。他總能將順德菜的靈魂融入世界食材之中，更能因地制宜，展現料理的當代詩意。2013 年，有幸引薦富哥到台灣（高雄），至今已拓展數間分店，富哥的魅力無窮！他曾與我攜手舉辦「白魚宴」，將藝術、美食、音樂與生活美學交織成章，共同傳遞「無隅」的真諦。

富哥是順德菜的傳承者，是味覺的詩人，更是我美食旅途中的知己，他的每一道菜，都是一場無隅的修行——誠懇、沉着、無聲，卻深深地打動人。

白魚 Bai Yu（本名林伯禧）

國際藝術家
比利時皇家美術學院 終身榮譽教授
米蘭布雷拉藝術學院 藝術最高榮譽文憑
盞斝爵老饕會 創會會長

推薦序二

無論是廣州、香港、澳門，以至世界各地，很多人對中華廚藝都有熱烈的追求及研究，而中國八大名菜之一的粵菜，更是廣泛流行。

我童年時代的同窗鄭錦富，人稱富哥，是名人坊高端粵菜食府的始創人，與他有六十餘年友緣。他紮根香港數十年，廚藝深厚，而且不斷研究創新粵菜，在傳統的粵菜上演繹宴席的私房菜名典，深受飲食名家及各地名人青睞。我曾應富哥之邀，參加國內名人坊開業之慶典，親身體驗富哥親自下廚的匠心，讓我和嘉賓圓桌品嘗多款名典佳餚，十分幸運，能夠體會不一樣的舌尖風味，滋味暖心。名人坊的成功，是富哥對食材選擇的特別要求，以及專業的加工，並對廚藝精益求精，一絲不苟。

此外，富哥對香港社會公益慈善一向不遺餘力，匠心奉獻，深受飲食界上流名人及平民百姓喜愛。這位「隱世廚神」曾在香港十四次榮登米芝蓮星級食府寶座之名，實至名歸，我作為老同學見證富哥獲得殊榮，深感自豪及驕傲。

我十分感謝飲食界名家、資深各界好友對《富哥名人坊》的鼎力支持，讓富哥的首本食譜著作與各界好友及忠實粉絲見面，讓《富哥名人坊》的中華美食家喻戶曉，以時俱進，享譽國內外地區，發揚光大。

吳標華

吳氏宗親會榮譽會長

序言一

中國人對廚藝的講究，是世界之冠，而廣東人對美食的追求，可說是全國之冠，中國八大菜系就包括粵菜。以廣府菜為代表的粵菜，崛起於明清，盛於民國，而香港粵菜在過去大半個世紀以來的強勢成長，成就了粵菜的發展和興盛，粵菜更是大多數香港人與生俱來的認知。

鄭錦富師傅，人稱富哥，與我相識不久便成為好朋友。富哥的廚藝，功力深厚，穩健而扎實是他的風格。富哥多年來對廚藝深入鑽研，對食材有嚴格及專業的要求，更不斷作出新嘗試，通過菜式演繹傳統粵菜，同時也與時共進加入創新思維，菜式出品皆為大師級之作。作為食評家，當人們請我推薦一家我喜歡的粵菜餐館，富哥的名人坊當為不二之選。

富哥作為粵菜的一級名廚，在大陸及台灣，以至香港地區各地捧場的粉絲甚眾，名人坊的成功，與富哥為人謙虛及親和力，絕對有關。多年來，不少出版社曾提出與富哥合作出版食譜書，但都被他婉拒，一天等到我這個與他性格相近的人出現，雙方一拍即合，並得到萬里機構的支持，於是成就了這本《富哥名人坊》。

我曾經出版過廿多本食譜書，慶幸長期得到讀者們的支持，近年我專注於研究和撰寫中國飲食文化，並致力完成了工程浩大的《香港志—飲食卷》。本以為 2021 年出版的《參乾寶貝》是我食譜書的收官之作，但當遇上魅力沒法擋的偶像廚師富哥，遂主動提出一起合作出版食譜書，富哥立即回答「好！我就是在等你。」

這是一本由我全程監製，富哥盡顯畢生功力，傾力演出的食譜書，書中收集了 52 道富哥的著名菜式，是一本廚藝愛好者值得用來閱讀、學習及收藏的好書，更是近年少見的高級粵菜食譜書。大師之作，不容錯過，我誠意向大家推薦。

在製作過程中，我與富哥攜手合作，感謝富哥對我工作時六親不認態度的容忍，同時富哥的菜式也使我增加了不少廚藝知識。

感謝好友張聰的 LEGLE 品牌，為本書拍攝時提供精美瓷器餐具，有了你，書中圖片更明亮美麗！

方曉嵐

飲食文化作家

2025 年春

方曉嵐

香港飲食文化作家、食評家、烹飪導師、報章專欄作家、香港中華廚藝學院大師級課程客座導師及論文評審。在香港出版了二十多本飲食文化食譜書及文字書，在台灣出版了四本食譜書。2014 年應歐洲出版社 Phaidon Press 的邀請，為國際食譜書系列撰寫英文版 *China the Cookbook*，於 2016 年出版及全球發行，該系列為各國主要圖書館永久收藏。*China the Cookbook* 現已翻譯成多國文字，包括德文、法文、西班牙文和簡體中文版，中文版《中国菜》被國家圖書館永久收藏。她曾擔任央視紀錄片「香港之味」總顧問，以及《香港志 — 飲食卷》總撰。

序言二

一味入魂，一生入藝

在芸芸廚藝之林中，有人技驚四座，有人情味動人，而我希望兩者皆備。

出於香港老派粵菜的薈萃之地，成長於百味人生的蒸騰煙火之中。從香港林百欣先生私廚到米芝蓮星級餐廳主理人，以「用心」為匙，打開一道道味蕾與人心之門。我的菜不僅僅是料理，更像是一場文化的延續，一則匠人的低語，一道藝術的光影。

「飛天雞」的創意、「燕窩釀鳳翼」的優雅，那不是炫技的浮華，而是對食材的敬畏、對傳統的回應、對客人的柔軟，更是我從不藏私，甘願以一書之形，把幾十年爐火純青的經驗與心法，毫無保留地傳遞給後輩。

在這本飲食書中，讀者不僅能學會做一道道菜，更能看見一道道「人」的道理，因為在我的廚藝世界裏，料理不只是舌尖上的記憶，更是內心的交流。

願你我翻開這書的同時，也像走進廚房的世界，感受那一盞燈下、一道菜中，用一生守護着那股真味傳承。

鄭錦富

2025 年夏

鄭錦富總廚

人稱富哥，米芝蓮星級大廚。富哥於 1973 年踏入廚藝界，曾於遠東交易所會所及私人俱樂部工作，並曾擔任名人家廚，及後成為名人坊高級粵菜食府總廚。憑着他豐富的經驗及獨特的烹調技藝，並多年來為各界名人提供私房菜佳饌，富哥被外界譽為「隱世廚神」。

富哥廚藝精湛，多次榮獲飲食界殊榮，包括 2009 年「飲食天王之隱世廚神大獎」，以及於 2010 年起連續六年帶領團隊獲得米芝蓮二星殊榮。在 2023 及 2024 年，富哥再闖高峰，由他主理的食肆榮獲米芝蓮一星食府之名，充分展現其烹調實力。富哥多次獲海外酒店及餐廳邀請獻技，也多次為香港旅發局到海外推廣香港美食，為海外食客炮製精彩絕倫的粵式美饌。

富哥與名人食客的美食之緣

富哥掌廚半世紀，其穩實的廚房功架、對食材的講究及巧思，主理具水準的手工粵菜，推崇者眾，名人食客無數，讓老饕們細嚐最頂級的粵菜料理。

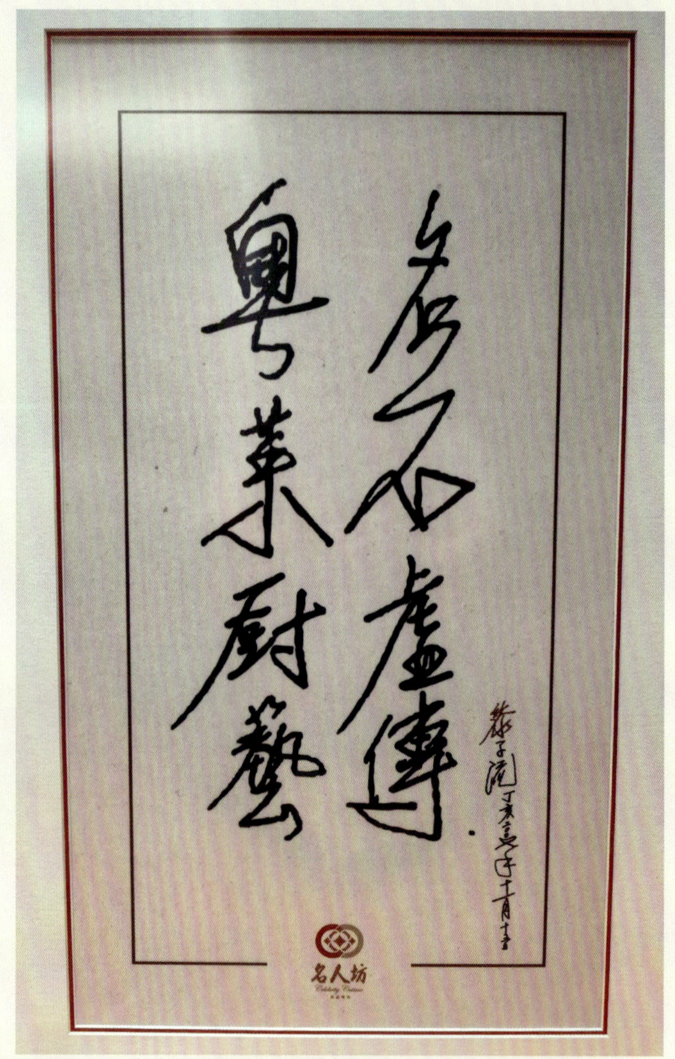

廣州市人民政府前市長黎子流，親筆為富哥題字。

富哥的廚藝獲得不少名人支持及讚譽，是富哥對粵菜不斷向前的推動力。

目錄

富哥精選佳餚

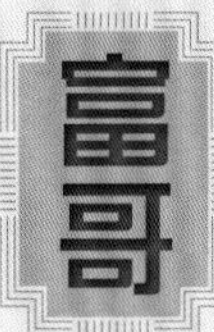

極品珍味

富哥 傳統經典菜

富哥 宴客佳饌

富哥 招牌飯麵

精選佳餚

掌廚半世紀，富哥的招牌菜如數家珍——
燕窩釀雞翼、豉椒炒肚尖、富哥鮑魚、花雕飛天雞……

富哥多年來做菜的宗旨——

不花巧、不嘩眾，用心鑽研與烹調，保留粵菜的韻味，

讓食客細嚐粵菜的真正味道。

燕窩釀鳳翼

這是一道九十年代創新的粵菜，在雞翼釀滿以高湯煨味的官燕，口感綿軟細滑，外皮炸得香脆，充滿雞皮油脂的香氣，是完美極致的契合，廣受食客歡迎。

一位用

材料

新鮮雞翼 1 隻

已浸發燕窩 20 克

上湯適量

生粉適量

麥芽糖水適量

醃料

鹽適量

做法

1. 光雞保留鳳翼（連彎位），去骨備用。
2. 燕窩用上湯煨透入味，隔乾，埋獻煮好，備用。
3. 雞翼釀入燕窩 20 克，以牙籤穿緊封口，以鹽醃味 3 小時。
4. 煮滾水，放入雞翼燙皮，隔水，淋上麥芽糖水（上皮水），掛於通風處待 2 小時以上至乾透。
5. 燒熱油至 120℃，放入雞翼浸熟，然後加油溫至 160℃至外皮香脆及金黃，吸乾油分，裝飾上碟。

富哥秘訣

選用官燕，即金絲燕燕窩，浸泡一晚後，再用高湯煨味，燕窩吸收高湯的味道及精華，與鮮香的雞肉非常搭配。

富哥鮑魚

鮑魚是海味之極品，經富哥精鍊的燉煮烹調，讓鮑魚散發獨特的鮮香風味，吸收鮮雞及瑤柱等湯汁精華，鮑魚口感綿軟，甘腴香甜，食味層次分明，是極品中之極品。

材料

乾鮑魚 3 公斤（5 斤）
鮮雞 2 隻（3.6 公斤）
瘦肉 1.2 公斤
腩排 1.2 公斤
雞腳 600 克
豬皮少許
江瑤柱 10 粒
葱適量

調味料

冰糖 20 克
老抽少許

做法

1. 乾鮑魚用水浸泡至軟身，約一星期（視乎鮑魚大小而定）。
2. 乾鮑魚隔水蒸至軟腍（約 8 小時，視乎鮑魚大小而定），盛起備用。
3. 鮮雞洗淨，斬件；瘦肉及腩排洗淨，切件；雞腳及豬皮洗淨；江瑤柱用水浸至軟身。
4. 燒熱鑊，加入葱爆香，放入雞件、瘦肉、腩排、雞腳及豬皮，油炸至香透，盛起。
5. 瓦鍋內墊上竹笪，排入雞件等材料，加入江瑤柱、鮑魚及調味料，燜 3 小時至鮑魚軟糯，隔去材料，過濾鮑汁，上碟伴菜享用。

富哥秘訣

- 富哥選用日本吉品鮑製作，香味濃郁，口感軟糯，呈溏心狀。
- 想提升鮑魚的鮮香味道，在燜煮鮑魚前，先用頂湯泡浸鮑魚，讓鮑魚吸收湯汁的鮮香，鮑香四溢。

招牌琵琶豆腐

因形似琵琶而得名，富哥將尋常的豆腐搭配鯪魚肉及金華火腿，油炸而成琵琶豆腐，外皮酥香、內裏嫩滑，豆腐混和各種鮮味，是晚飯必選菜式。

材料

長豆腐 6 件

鯪魚肉 200 克（5 兩）

雞蛋白（蛋清）約 5 個

金華火腿 30 克

乾葱 38 克

芫荽 38 克

已浸發陳皮絲 20 克

澄麵少許

調味料

鹽 10 克

糖 10 克

生粉 60 克

做法

1. 豆腐切成細粒狀，隔乾水分，待約 30 分鐘。
2. 金華火腿切細末；乾葱及芫荽切碎，備用。
3. 鯪魚肉、雞蛋白、金華火腿末、乾葱碎、芫荽碎、陳皮絲及調味料搓勻，拌入豆腐粒。
4. 預備匙羹，放入豆腐漿蒸熟定形（約蒸 6 分鐘），取出，待涼備用。
5. 輕輕掏出已定形的豆腐塊，撲上澄麵，放入熱油炸至金黃色，上碟享用。

富哥秘訣

餡料可選配蝦漿、葱花等，與豆腐拼發另一種鮮香滋味。

豉椒炒肚尖

挑選豬肚最嫩滑的部位，切得薄透，是刀工精妙之處，配合精準的火候，嚐到的是豬肚尖爽嫩的口感，將粵菜鮮味的精髓表露無遺，是珍貴的功夫菜。

材料

豬肚 4 至 6 個（約 160 克）

三色甜椒及洋葱 160 克

豆椒 2 茶匙

薑蓉、蒜蓉及乾葱蓉各少許

醃料

水蟹 1 隻（取蟹水）

糖 4 克

生粉 12 克

調味料

鹽、糖、老抽各少許

做法

1. 豬肚用粗鹽擦淨內外，洗淨，切出肚尖最尖端部位，去除筋膜及油脂，切成薄片狀。
2. 豬肚尖用醃料拌勻，醃味 1 小時。
3. 三色甜椒去內瓤，洗淨，切塊；洋葱去外衣，洗淨，切塊。
4. 燒熱油鑊，下薑蓉、蒜蓉、乾葱蓉及豆豉爆香，放入三色甜椒、洋葱及豬肚尖，以中火快手炒香，灑入調味料拌勻，即可上碟。

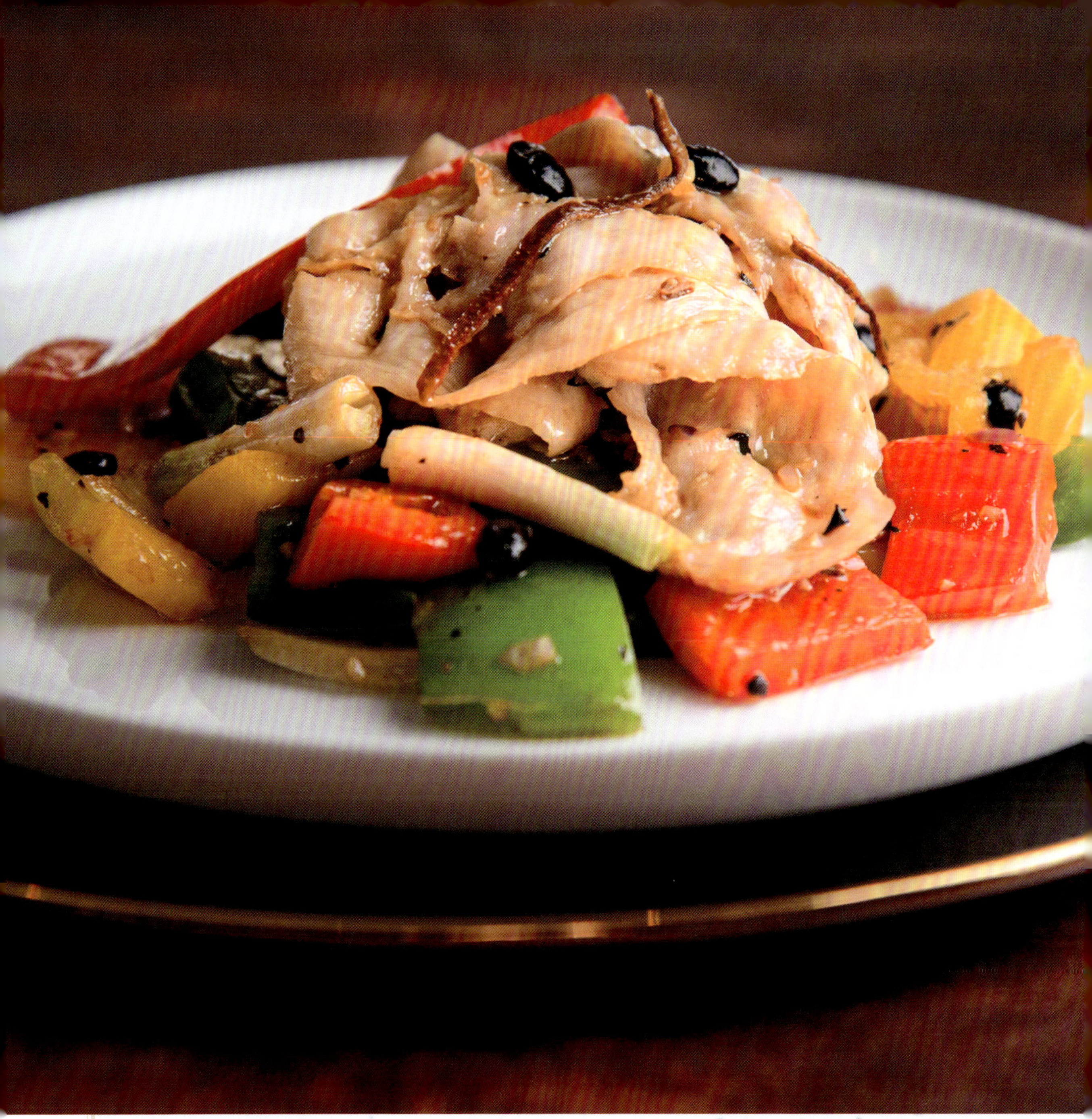

富哥秘訣

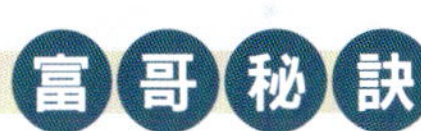

- 豬肚尖佔整個豬肚只有六分之一，所以烹調一道豬肚尖菜式，必須最少選取四個豬肚。
- 火候要拿捏準確，要用猛火爆炒，才能嚐到豬肚尖的爽滑質感。

百花釀蟹箝

一位用

外酥內嫩的百花釀蟹箝，蝦膠及蟹肉完美地融合，彈牙爽口，每口是海鮮的鮮甜滋味，足證大廚拿捏食材的卓見，是粵菜經典的菜式。

材料

花蟹蟹箝 280 克

蝦膠 60 克

方包 2 片

蝦膠材料

中蝦 600 克

肥肉粒 20 克

蛋白 1 個

鹽、糖及生粉各適量

蝦膠做法

1. 中蝦去殼，挑去蝦腸，用粗鹽抓洗，再用水洗淨，用乾布壓乾水分。
2. 蝦肉放於砧板，用菜刀刀身壓扁蝦肉，再用刀背粗剁成蝦蓉。
3. 蝦蓉放入攪拌盤，加入蛋白，用手順一方向攪拌至起膠，再加入肥肉粒、鹽、糖及生粉等調味料拌勻，用手抓起蝦蓉並撻回盤內，反複數次至呈膠狀，即成蝦膠。

做法

1. 蟹箝沖洗乾淨，去掉碎殼，瀝乾水分。
2. 方包切成粒，待乾透，成麵包糠。
3. 蟹箝均勻地抹上蝦膠，整個裹好，黏上麵包糠，壓實。
4. 燒熱油，放入蟹箝以中火慢浸，見蟹箝浮起及熟透，炸至金黃色，上碟裝飾即可。

- 整個花蟹箝裹上蝦膠，嚐到兩種不同的口感層次。。
- 麵包糠以麵包切粒製成，不同於粉狀的麵包糠，脆口感覺更加豐富。

香檳汁脆皮刺參

一位用

刺參，一般以燜煮方式饗客。富哥來個反傳統，將蟹肉及洋葱釀入刺參，再炸脆擺盤，以忌廉及上湯做成西式的香檳汁，令人耳目一新。

材料

北海道刺參 1 支（乾品重量 600 克含 90-100 支）

薑片、葱段及蒜子各少許

蟹肉 10 克

洋葱絲 20 克

蒜蓉少許

忌廉、上湯及牛油各少許

麵撈適量

蛋白漿

蛋白 2 個

粟粉 40 克

生粉 20 克

香檳汁

忌廉、上湯、牛油及香檳酒各少許

青紅甜椒粒少許（裝飾用）

炸脆米少許（裝飾用）

做法

1. 發好刺參至軟身，備用。
2. 薑片、葱段及蒜子用上湯煨刺參入味，瀝乾備用。
3. 燒熱油，放入蒜蓉、洋葱絲及蟹肉拌炒，倒入忌廉、上湯及牛油煮好，加入麵撈拌至濃稠，煮至蟹肉剛好不太軟身，盛起。
4. 將蟹肉洋葱釀入刺參，蘸上拌勻的蛋白漿。
5. 用熱油把刺參炸脆至金黃色，上碟。
6. 將香檳汁材料埋薄獻，加入青紅甜椒粒，淋於碟上，放上刺參，最後灑入脆米即可。

富哥秘訣

- 不要將蟹肉煮得太軟腍，釀入海參後，口感不太好。
- 麵撈以牛油及麵粉煮成濃稠的麵糊，起着調整濃稠度的作用，多用於西式菜餚。

花雕飛天雞

飛天雞是富哥得意之作，加入花雕酒以舊式電飯煲焗製，雞肉滲滿酒香，操作簡便，是大廚對美食的熱誠及巧思。

材料

活雞 1 隻（約 1.8 公斤）

花雕酒 200 克

青葱段適量

醃料

鹽 16.8 克

做法

1. 新鮮雞洗淨，去內臟，抹乾水分，加入鹽醃 4 小時。
2. 在電飯煲內膽鋪上青葱段，放入鮮雞，雞肚朝上放好，澆入花雕酒，按掣焗煮 40 分鐘至雞肉熟透，上碟享用。

富哥秘訣

- 醃製雞隻厚肉部位需要多加鹽分，一隻光雞淨重 1.8 公斤（約 2.12 斤），每斤雞約需 5.6 克鹽。
- 青葱段鋪墊放電飯煲內膽底部或瓦煲底，雞肚向上，可保存雞汁豐腴，而且雞胸肉容易熟透。

鵝掌扣鮑魚

一位用

鮑魚養肝強身，以十二頭或十三頭日本吉品鮑入饌，頂級之食材配上大師級烹調技藝，溏心綿軟，味道細膩，令食客一試難忘。

材料

鮑魚 1 隻

鵝掌 1 隻

菜遠 2 棵

炆鵝掌料（可炆鵝掌 100 隻）

薑 8 片

葱 8 條

蒜肉 150 克

乾葱 75 克

肥肉 1.2 公斤

金華火腿骨 900 克

鵝掌調味料

老抽少許

水適量（可蓋過鵝掌）

蠔油 225 克

乾鮑材料

乾鮑魚 3 公斤（5 斤）

鮮雞 2 隻（3.6 公斤）

瘦肉 1.2 公斤

腩排 1.2 公斤

雞腳 600 克

豬皮少許

江瑤柱 10 粒

葱適量

乾鮑調味料

冰糖 20 克

老抽少許

乾鮑做法

1. 乾鮑魚用水浸泡至軟身，約一星期（視乎鮑魚大小而定）。
2. 乾鮑魚隔水蒸至軟腍，約 8 小時（視乎鮑魚大小而定），盛起。
3. 鮮雞洗淨，斬件；瘦肉及腩排洗淨，切件；雞腳及豬皮洗淨；江瑤柱浸至軟身。
4. 燒熱鑊，用葱爆香，放入雞件、瘦肉、腩排、雞腳及豬皮，炸至香，盛起。
5. 瓦鍋內墊上竹笪，排入雞件等材料，加入江瑤柱、鮑魚及調味料，燜 3 小時至鮑魚軟糯，隔去材料，過濾鮑汁，備用。

鵝掌做法

1. 鵝掌洗淨，用少許老抽上色。
2. 燒熱油，冒少許白煙，放入鵝掌炸 30 秒至淺金黃色，盛起。
3. 炒煮薑、葱、蒜肉及乾葱，加入肥肉及金華火腿骨，以調味料燜約 1 小時 15 分鐘，加蓋再待 1.5 小時至軟腍。
4. 上碟時配上吉品鮑及灼菜遠，趁熱品嘗。

富哥秘訣

- 選用波蘭出產的鵝掌，肉厚體大，軟滑可口，極之美味。
- 鵝掌上色用的老抽，不要選太深色，以免鵝掌色澤太黑，影響賣相。
- 鵝掌需要以肥肉燜煮，讓鵝掌增添香氣。

竹笙釀官燕

一位用

這是一道精緻的高級粵菜，軟滑濃郁的官燕，搭配爽脆鮮香的竹笙，完全吸收上湯的精華，清爽、香滑、味濃，給食客帶來難忘的食味回憶。

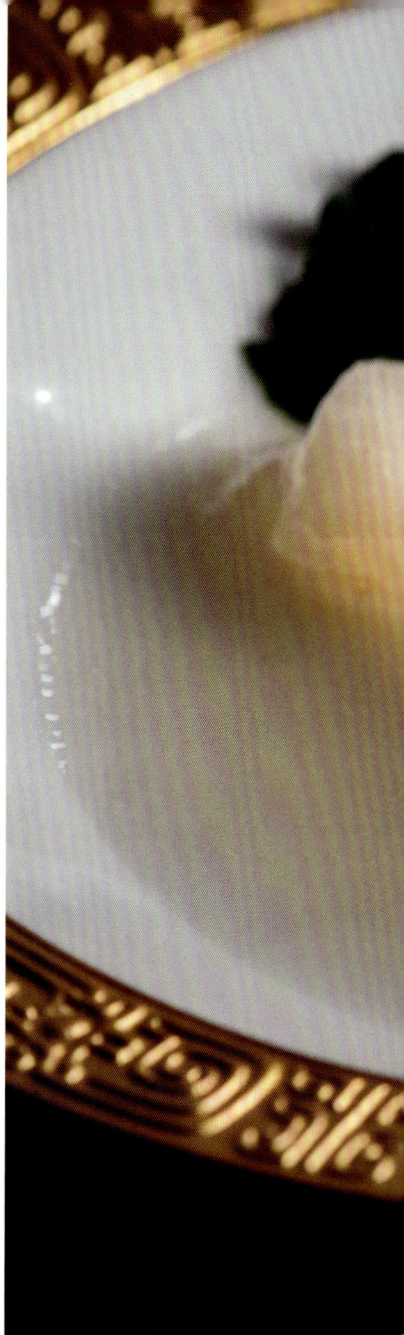

材料

竹笙 1 條（略粗，7 至 8 厘米長）

官燕 38 克

菜遠 2 棵

上湯適量（可蓋過燕窩）

熟金華火腿絲少許

做法

1. 竹笙用水浸泡 30 分鐘至軟身，去掉頭尾兩端，飛水，以上湯煨味，備用。
2. 官燕處理好，加入上湯蒸 12 分鐘，隔水，埋獻煮好，備用。
3. 將官燕釀入竹笙，上碟，伴菜遠，最後淋上湯獻，以熟金華火腿絲裝飾即可。

富哥秘訣

竹笙及燕窩吸收上湯精華，達至味道融和，不妨花點時間炮製上湯，以老雞及金華火腿等燉煮上湯，自然能嚐到與別不同的食味。

炒桂花魚翅

民國時期經典的廣府粵菜，雞蛋炒成桂花形狀，配以蟹肉、魚翅及銀芽，爽口有嚼勁，層次豐富，着重大廚拿捏火候的技巧，以及精巧的手藝，體現富哥的匠心獨運。

材料

魚翅 120 克（已泡發）
蟹肉 38 克
銀芽 58 克
叉燒絲 20 克
金華火腿絲少許
雞蛋 2 個（大）
韭黃少許
上湯適量（煨魚翅用）

調味料

鹽少許
生粉水少許
紹酒、麻油各適量

做法

1. 魚翅洗淨，蒸至軟身，以薑葱煨味去腥，加入上湯燉煮，盛起備用。
2. 銀芽以上湯煮至 6 成熟，盛起。雞蛋灑入少許鹽拌勻。
3. 燒紅鍋，放入雞蛋快炒至桂花粒，加入魚翅及蟹肉炒香，下叉燒絲、銀芽及韭黃炒勻，灑入紹酒、麻油及生粉水拌炒，最後綴以金華火腿絲裝飾即成。

富哥秘訣

處理魚翅必須有耐性，它是整道菜的靈魂，魚翅煨透入味，呈現柔嫩滑軟的口感，令整道經典粵菜達至更高水平。

脆皮炸子雞

雞皮薄脆，光澤勻稱，雞肉鮮嫩多汁，皮下脂肪入口即溶，油香豐足，呈現富哥深厚的功力，是老饕食客必點的菜式。

材料

光雞 1.8 公斤

鹽 15 克

上皮水

白醋 200 克

麥芽糖 50 克

大紅浙醋 5 克

雙蒸酒 5 克

做法

1. 預備上皮水：白醋、麥芽糖及大紅浙醋放於碗內，座於熱水至糖溶，待涼，拌入雙蒸酒，備用。
2. 雞洗淨，去掉內臟，沖淨，抹乾水分。
3. 雞內外用鹽抹勻，醃味 4 小時。
4. 燒滾水，放入雞輕燙外皮，待乾，再抹上皮水，掛起風乾待 6 小時至完全乾透。
5. 中火燒熱油至 120℃，放入雞浸 10 分鐘，調大至猛油 160℃，炸至皮脆即可，快手盛起，斬件上碟享用。

富哥秘訣

- 富哥強調最重要是選用新鮮的三黃雞製成炸子雞。
- 要做到皮脆肉嫩的效果，油溫的控制非常重要，先用 120℃浸炸至肉熟嫩滑，再調大至 160℃炸至外皮酥脆。

紅燒佛跳牆

一位用

佛跳牆是福建的名菜，正宗佛跳牆由鮑參翅肚等上乘材料製成，用料考究。富哥將各種高級食材完美烹調，成為高級的滋補佳品。

材料

吉品鮑 1 隻（35 頭）

已浸發花膠（鴨包肚）40 克

海參 60 克

魚翅 20 克

鮑魚汁上湯 80 克

乾鮑材料

乾鮑魚 3 公斤（5 斤）

鮮雞 2 隻（3.6 公斤）

瘦肉 1.2 公斤

腩排 1.2 公斤

雞腳 600 克

豬皮少許

江瑤柱 10 粒

葱適量

乾鮑做法

- 參考本書第 35 頁吉品鮑的做法。

花膠做法

- 花膠用滾水浸發至軟腍（需時數天），以薑片、葱、炸蒜子及二湯煨味去腥，備用。

海參做法

1. 海參先用水浸一至二天，再用滾水加蓋焗一晚。
2. 剪開海參腹部，清洗腸砂，再焗至軟腍，浸泡冷水。
3. 將薑片、蒜子、葱段及二湯煨海參（凍浸），以上湯用慢火燜海參，備用。

魚翅做法

- 魚翅浸發好，洗淨，蒸至軟身，以薑葱煨味去腥，加入上湯燉煮入味，盛起備用。

綜合做法

- 各款食材燜煮妥當，煮滾鮑魚汁上湯，放入鮑參翅肚略煮，埋獻，上桌享用。

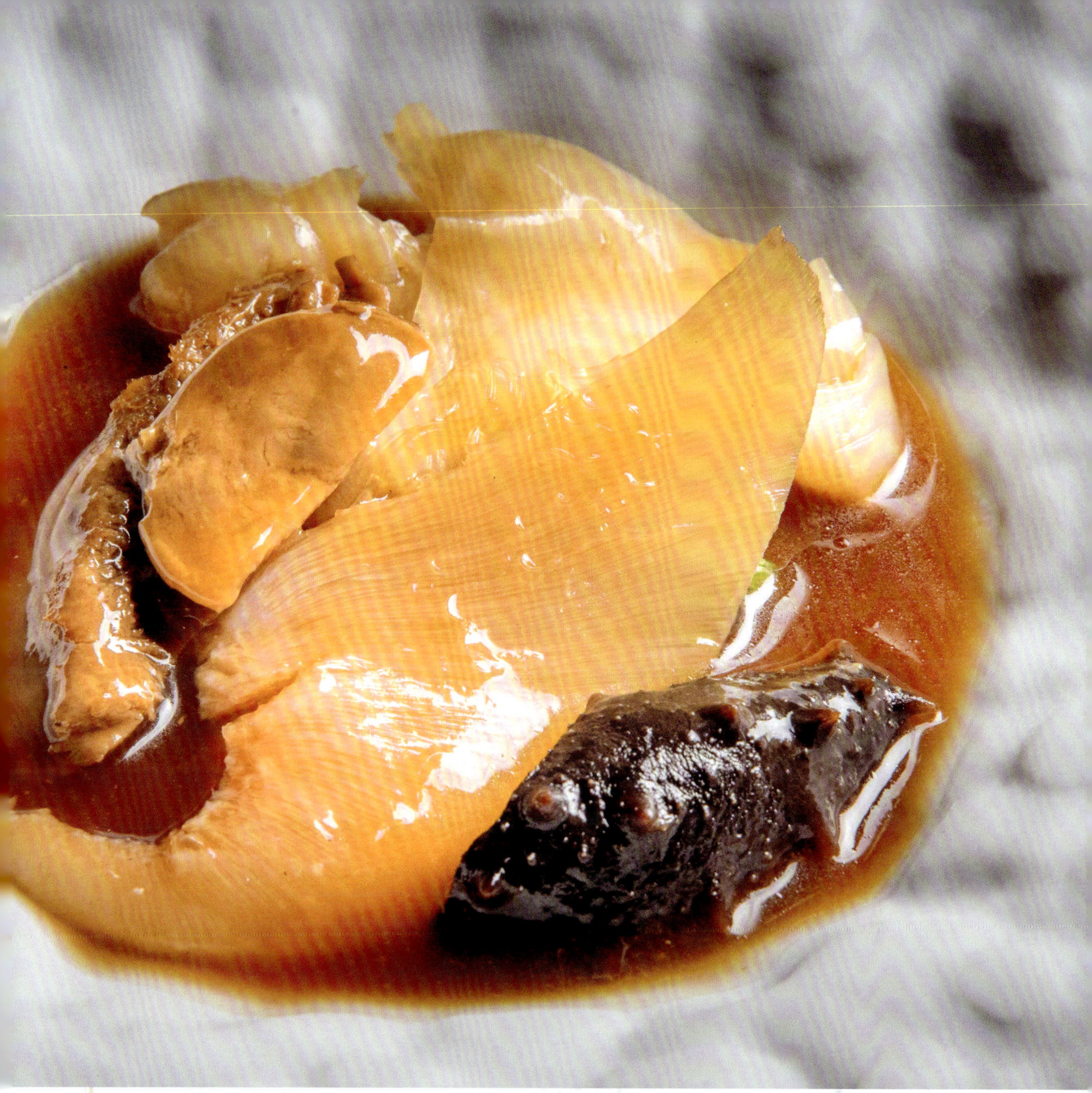

富哥秘訣

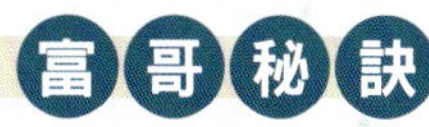

每款食材的浸發及燜煮時間需要拿捏準確，才能烹調一道完美的紅燒佛跳牆，絕對是考驗廚師的功力。

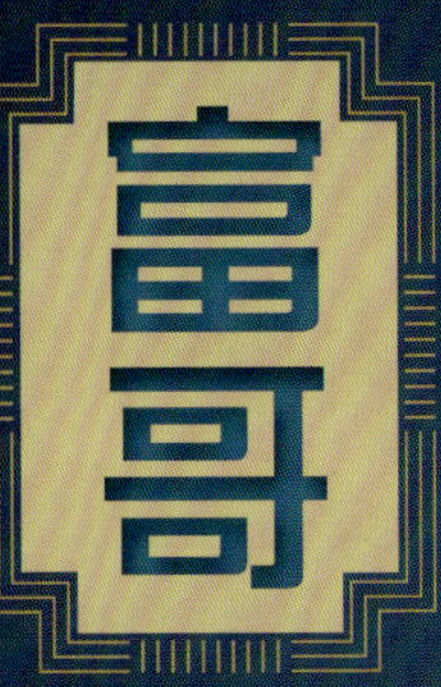

極品珍味

花膠、海參、魚翅、燕窩——
珍貴的海味及滋補食材，是粵菜宴客的精緻佳饌。

富哥着重選料配材，他深入了解不同地域出產的材料，
嚴選上佳的海味及乾貨寶物，施以入廚魔法，
炮製珍饈百味，為食客帶來夢幻般的嚐味喜悅。

旨人乾撈翅

一位用

晶瑩剔透的翅針，配上濃郁入味的上湯，乾撈翅最是講究大廚烹調的火候及功夫，掌握魚翅煨味的時間控制，是一道盡顯心思的菜式。

材料

魚翅 180 克

金華火腿絲少許

薑片及葱段各適量

煨魚翅料

紹興酒少許

濃雞湯 80 克

琉璃獻適量

伴吃

銀芽及韭黃各少許

上湯 1 碗

做法

1. 魚翅浸發好，洗淨，蒸至軟身，以薑葱煨味去腥。
2. 燒熱鑊，灒紹興酒，加入濃雞湯煨魚翅至入味，上碟。最後以琉璃獻拌成獻汁，淋於魚翅上，以金華火腿絲裝飾。
3. 享用時，魚翅、銀芽、韭黃及上湯一併奉客。

要控制魚翅煨、煮、燉的火候及時間，以免令魚翅煮得太糊，欠爽滑針翅的口感。

雲腿菜膽燉鮑翅

一位用

將鮑翅、菜膽及雲南火腿放入燉鍋，加入上湯慢燉而成，清香、味鮮，齒頰留香，是一道色香味俱全的傳統粵菜。

材料

鮑翅 80 克

菜膽 2 棵

雲南火腿 1 片

上湯 240 克

薑片及葱段各少許

上湯料

老雞 1 隻

瘦肉 480 克

豬腿骨 160 克

牛肉 140 克

金華火腿 140 克

水 4.8 公升

做法

1. 上湯料處理乾淨，放入大鍋內煲煮 2-3 小時，最後得上湯約 2.4 公升，取 240 克上湯用。
2. 菜膽飛水，用上湯煨入味，瀝乾水分。
3. 鮑翅洗淨，用竹笪夾好，蒸約 6 小時至針翅軟腍。
4. 鮑翅用薑片及葱段煨好，去腥味。
5. 將鮑翅、雲腿及菜膽放入燉盅，加入上湯燉約 2 小時即成。

建議用竹笪夾着鮑翅蒸軟，能保留原件大片鮑翅，賣相吸睛。

酸湯煎翅

香煎魚翅一般較少見，富哥以創意炮製，煎至香脆的魚翅配上酸甜味美的酸湯，非常吸引，翅針脆中帶滑嫩，滿口鮮味，享用後暖意盈盈。

材料

發好牙揀翅 1 隻

上湯適量

薑片及葱段各少許

酸湯材料

西芹、洋葱、青紅甜椒、紅蘿蔔、

番茄、鹹菜共 1.8 公斤

鮮魚 1.6 公斤

瘦肉 1.6 公斤

花椒粒少許

白米醋少許

水 7.2 公升

做法

1. 酸湯材料處理妥當，全部放進煲內，煮至水餘下 3.6 公升，備用。
2. 牙揀翅浸發好，洗淨，放於竹笪夾着，蒸至軟身，以薑片及葱段煨味去腥。
3. 牙揀翅用上湯煨好入味，備用。
4. 燒熱鑊，放入牙揀翅煎香兩面，盛於碟上。
5. 取適量酸湯，以薄獻埋獻，淋在煎翅即可享用。

富哥秘訣

- **牙揀翅性價比高，是背鰭位置的魚翅，翅針幼密，口感滑溜，能夠吸收酸湯的精華。**
- **煎翅要花點耐性，不能性急，因魚翅濕的話容易黏着鑊，要待鑊燒熱才放入魚翅。**

紅燒花膠王

一斤四隻的印度鰵魚膠，體大，質感豐厚，膠質豐富，是花膠界的上上之品。上桌奉客，碩大的花膠王絕對成為全場焦點，打卡聲不絕於耳。

材料

印度鰵魚膠 1 隻（4 頭）

上湯或鮑魚汁 320 克

薑片、葱段及蒜子各少許

上湯 1.14 公斤

生菜 6 棵

做法

1. 鰵魚膠浸發妥當，放入滾水加蓋待至軟腍。
2. 鰵魚膠用薑片、葱段、蒜子及上湯煨味去腥，再用上湯煨至入味及腍滑，上碟。
3. 煮滾上湯或鮑魚汁，埋獻，淋在花膠上，伴灼生菜享用。

富哥秘訣

鰵魚膠產於印尼或巴基斯坦，質感厚實，屬高級食材，大廚必須掌握浸發及煨煮的時間，才不致浪費矜貴的食材。

鮑汁扣日本刺參

一位用

這是傳統粵式名菜，富哥精選北海道刺參，刺粗而密，肉質爽滑；而鮑汁更是整道菜式的靈魂，如何做好鮑汁，就要看大師的功力了。

材料

刺參 1 支

鮑魚汁少許

上湯少許

菜遠 2 棵

煨料

薑片、蒜子、葱段及上湯

各適量

做法

1. 刺參先用水浸一至二天，再用滾水加蓋焗一晚。
2. 剪開刺參腹部，清洗腸砂，焗至軟腍，浸泡冷水，備用。
3. 將薑片、蒜子、葱段及上湯煨海參（凍浸），以上湯用慢火燜刺參，備用。
4. 煮滾鮑魚汁，放入刺參略煮，埋獻，伴菜遠上桌享用。

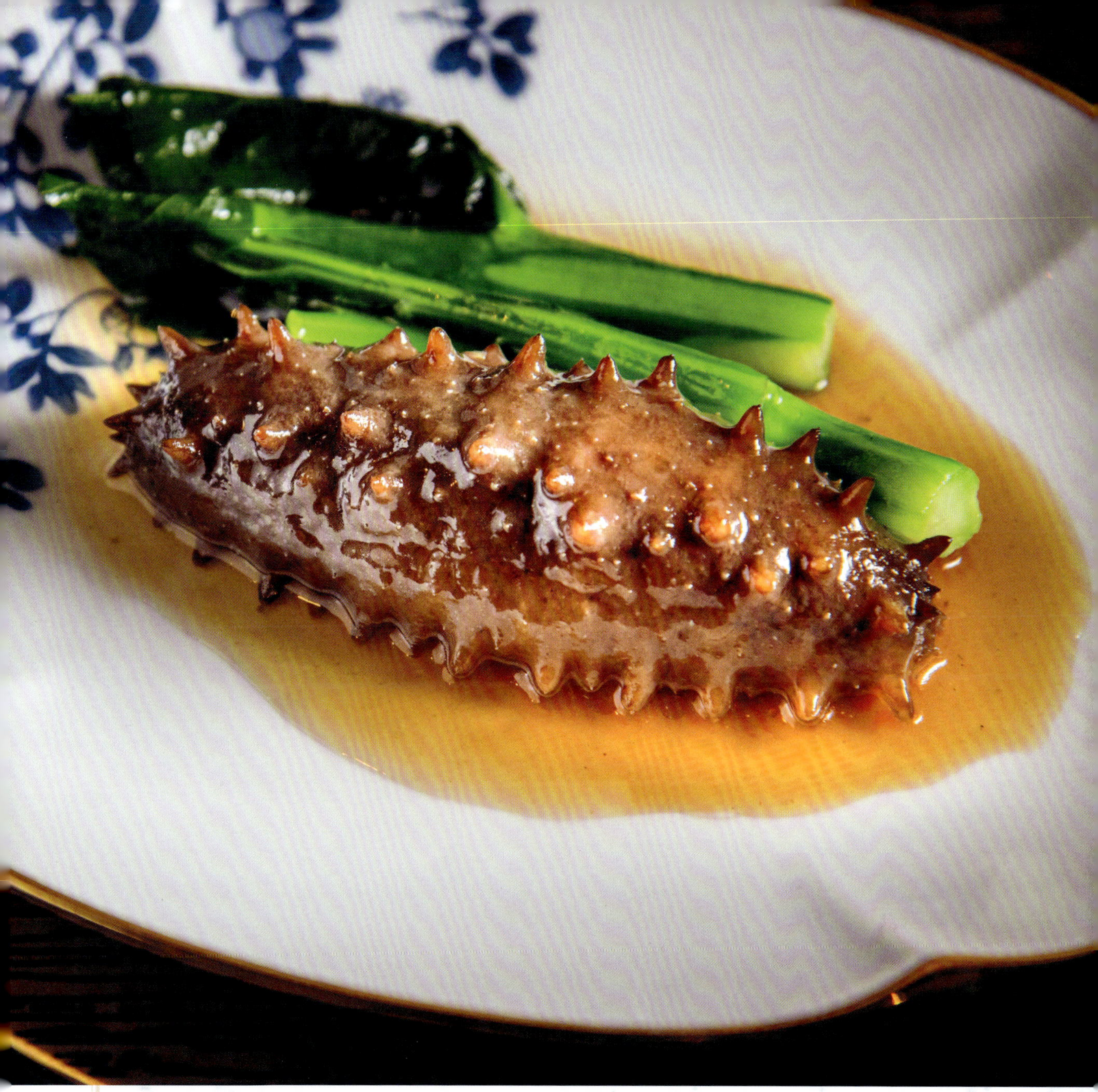

富哥秘訣

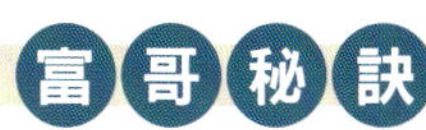

挑選日本北海道刺參，長年在清澈、水溫低的海水中生長，口感爽滑，在天然環境中的刺參，不但珍貴，而且營養價值非常高。

雞汁燴官燕

一位用

名貴的官燕，搭配由原隻雞清燉的濃雞汁精華，味道精醇濃郁，有別於一般上湯，足證富哥對菜式的堅持，時刻為食客炮製完美的名饌。

材料

官燕 80 克

光雞 1 隻（1.8 公斤）

上湯適量

金箔少許

做法

1. 官燕浸泡一晚，用上湯煨透入味及軟脸，隔乾水分，備用。
2. 光雞洗淨，去除內臟，沖淨，原隻隔水蒸 3 小時，取濃雞汁 80 克。
3. 濃雞汁埋稀獻，上碟，排上已煨味官燕，最後加上金箔裝飾即成。

以原隻雞燉煮而成的濃雞汁，過程中不含一滴水分，萃取雞汁精華，搭配官燕享用，品嘗到官燕的絲滑及雞汁的醇和。

燕窩鷓鴣粥

一位用

經典的懷舊手工菜，製法繁複，將鷓鴣起肉，配以參薯蓉及上湯等推成沒有米粒的鷓鴣粥，並綴以官燕，嚐到鷓鴣的鮮甜及官燕的淡香，還感受大廚對粵菜的傳承精神。

材料

官燕 80 克

鷓鴣蓉 25 克

參薯蓉 60 克

金華火腿蓉少許

上湯 80 克

做法

1. 官燕浸泡一晚，用上湯煨透入味及軟身，隔乾水分，備用。
2. 鷓鴣洗淨，去除內臟，沖淨，起肉留用。
3. 鷓鴣肉去筋膜，用刀剁成細蓉狀，備用。
4. 燒熱上湯，放入鷓鴣蓉及參薯蓉煮成粥，埋獻，煮成濃稠狀。
5. 鷓鴣粥分放碟內，排上官燕，灑上金華火腿蓉享用。

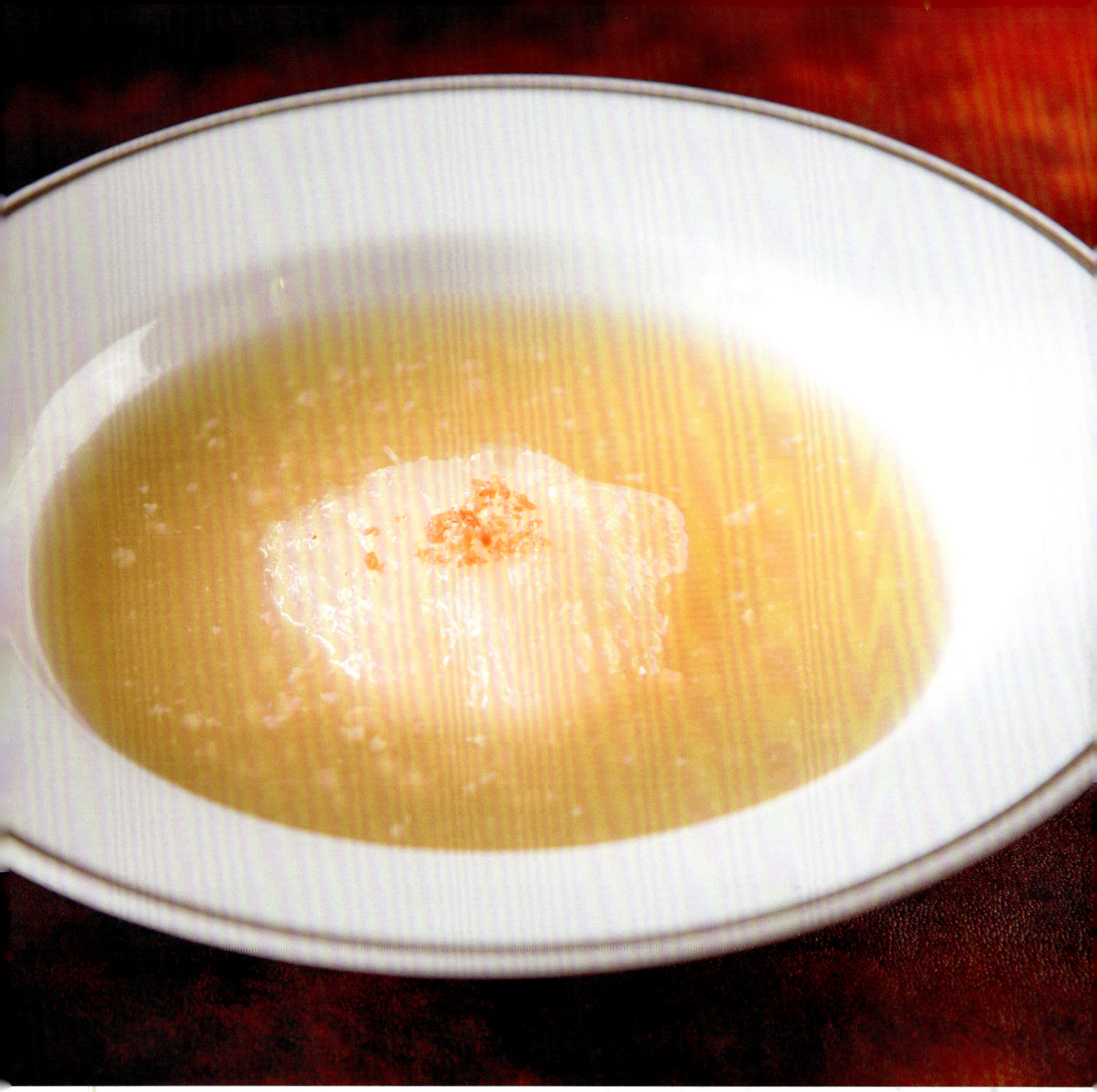

富哥秘訣

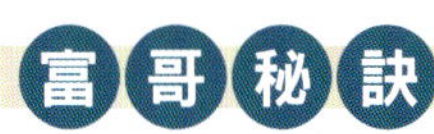

鷓鴣肉必須去除筋膜，細剁研磨成細蓉，與其他材料煮成粥狀，口感細緻綿密，是大戶人家小孩的補身食法。

冰花燉官燕

一位用

絲絲滑嫩的上等官燕，與清甜的冰糖燉製，兩者融和，伴以花奶及椰漿，是一頓豐富粵菜的完美終結。來一客矜貴的粵式甜品吧！

材料

官燕 80 克

冰糖少許

清水 80 克

花奶及椰漿各少許

金箔少許

做法

1. 官燕浸泡一晚，瀝乾水分。
2. 燉盅內放入官燕、冰糖及清水，燉至冰糖完全溶掉，隔起官燕，置於碗內。
3. 最後倒入花奶及椰漿，綴以金箔享用。

富哥秘訣

官燕與冰糖燉至融和，燕窩吸收冰糖的精華，此時隔起官燕，以免水分太多。品嘗絲絲柔滑的官燕，是潤肺補身的粵式甜品。

富哥

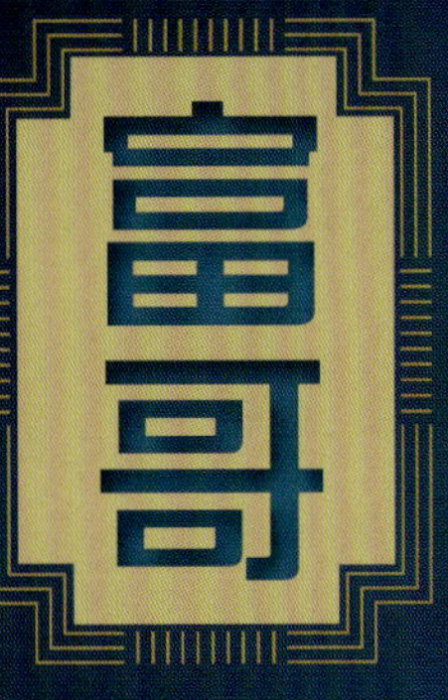

傳統經典菜

富哥一直為名人食客炮製高級的懷舊菜式 ——

福祿百寶鴨、香煎鳳尾蝦、江南百花酥雞、焗釀鮮蟹蓋……

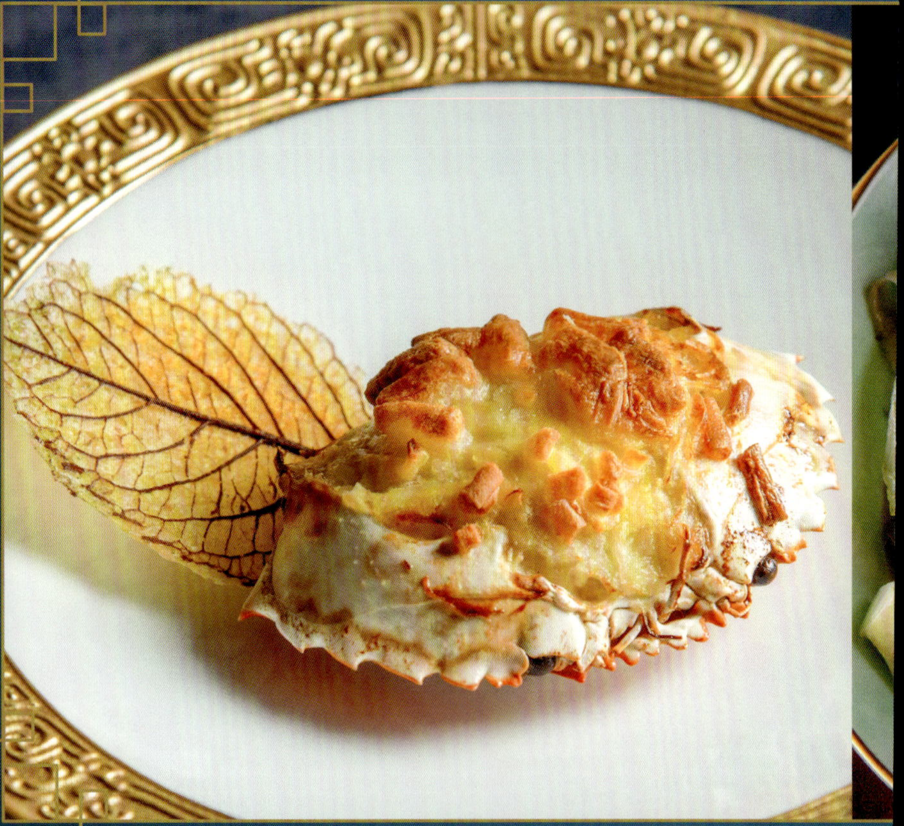

富哥是烹調懷舊菜的老行尊，
每時每刻以傳承經典懷舊菜為重。
他的傳統菜保留了昔日情懷，品嘗每啖滋味，
能讓食客體會失傳已久的老香港味道。

江南百花酥雞

這是一道矜貴的手工菜式，靈感來自上世紀二十年代廣州四大酒家之一文園酒家的江南百花雞。將蝦膠釀在雞皮中，同時嚐到爽口的蝦膠及脆嫩的雞皮。

材料

雞 1 隻

蝦膠 500 克

蝦膠材料

海中蝦 1.2 公斤

肥肉粒 20 克

蛋白 1 個

鹽、糖各 4 克

生粉 12 克

上皮水

白醋 200 克

麥芽糖 50 克

大紅浙醋 5 克

雙蒸酒 5 克

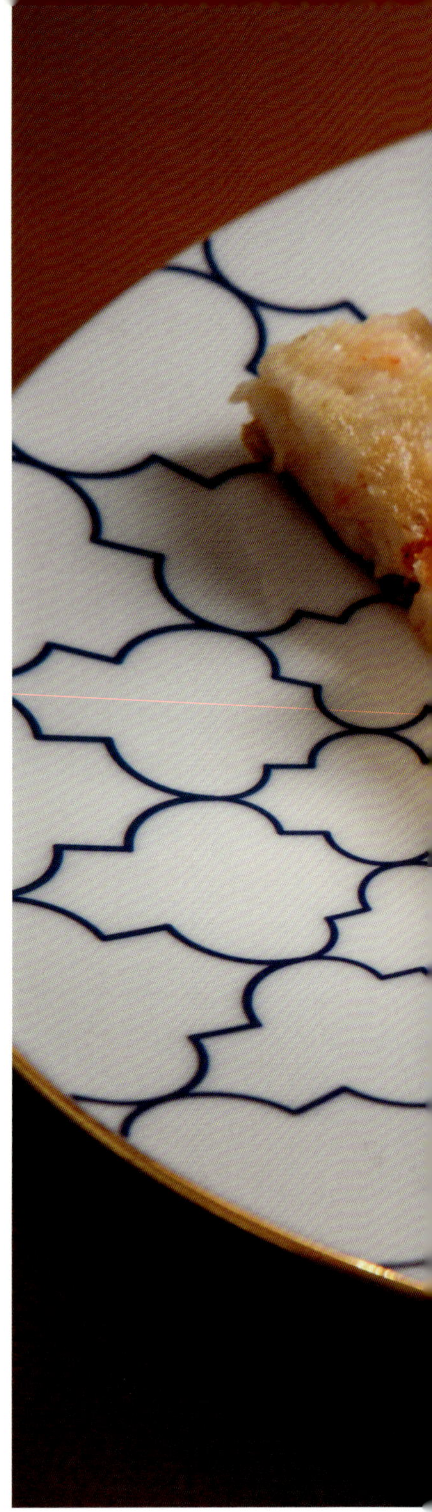

蝦膠做法

- 參考本書第 24 頁製作蝦膠的做法。

上皮水做法

- 白醋、麥芽糖及大紅浙醋混和，座於熱水至麥芽糖完全溶掉，待冷，拌入雙蒸酒拌勻。

做法

1. 光雞洗淨，去掉雞肉及雞骨部分，保留原隻雞皮，細心地褪出雞肉及骨以免皮層散爛。
2. 將雞皮用針固定在竹籬上，澆上熱水，反複數次，待雞皮收縮及定型。
3. 雞皮塗抹上皮水，吊起，待乾透。
4. 雞皮內抹上生粉，均勻地釀入蝦膠，放入熱油炸熟至金黃色，切件享用。

雞皮內必須塗勻生粉，讓蝦膠緊緊黏着雞皮，油炸後不易脫落。

富哥咕嚕肉

富哥的招牌菜之一，也是傳統的粵菜美饌。富哥特選豬梅頭爽肉炮製，滾上薄薄的炸粉，每口咕嚕肉香脆爽口，配上酸甜惹味的醬汁，成為大人及小朋友皆喜歡的經典菜。

材料

豬梅頭爽肉 113 克

青紅甜椒、洋葱 160 克

新鮮菠蘿 6 件

糖醋汁適量

料頭

薑蓉少許

蒜蓉 2 茶匙

葱度少許

做法

1. 豬梅頭爽肉切成厚片（長 3cm、闊 1cm、厚 3mm），沖淨，吸乾水分。
2. 青紅甜椒去籽，切角，飛水；洋葱及菠蘿切角備用。
3. 梅頭爽肉蘸上炸粉稀漿，再撲上生粉，放入熱油炸脆，盛起，再翻炸一次，隔油盛起。
4. 燒熱油，放入料頭及洋葱爆香，倒入糖醋汁煮熱，下生粉水勾芡，放入炸梅頭爽肉拌匀，最後下青紅甜椒及菠蘿炒匀，上碟。

富哥秘訣

- 每頭豬只有梅頭爽肉 450 克，更顯矜貴；富哥特意將豬爽肉切成片狀，只需短時間炸透，質感彈牙香脆，吃出不一樣的咕嚕肉。
- 糖醋汁製法：糖、片糖、白醋、茄汁、OK 汁、喼汁、番茄、西芹、芫荽、青紅甜椒及紅辣椒煮至雜菜軟腍，隔渣取汁。

焗釀鮮蟹蓋

一位用

蟹蓋釀滿一梳梳新鮮蟹肉，夾雜洋葱及薯蓉，啖啖鮮甜的蟹肉香，蟹味濃郁，外形飽滿金黃，是粵菜著名的巧手美饌，你又怎能抗拒它的鮮味誘惑？

材料

花蟹肉 32 克

洋葱絲 48 克

牛油少許

忌廉奶 40 克

上湯 40 克

薯蓉少許

蛋黃 1 個

芝士碎少許

蟹蓋 1 個

做法

1. 燒熱油，放入花蟹肉 10 克及洋葱絲炒香，加入牛油、忌廉奶及上湯拌勻，盛起。
2. 將花蟹肉 22 克釀於蟹蓋，填入蟹肉洋葱餡料。
3. 蟹蓋面塗抹薯蓉，輕掃蛋黃漿，撒上芝士碎，放入焗爐以 220℃焗 12 分鐘，至表面金黃即成。

富哥秘訣

- 將 22 克鮮蟹肉先釀入蟹蓋內，再鋪上蟹肉洋葱餡料，吃到的是滿滿的蟹香，分量豐富。
- 炒煮蟹肉洋葱餡料時，宜減少水分及調味分量，以免味道太稀或太濃，蓋過蟹肉之鮮香。

香煎鳳尾蝦

一位用

鳳尾蝦是粵式傳統名菜，潔白的蝦肉，配上鮮紅的蝦尾，狀似鳳尾而得名。大蝦肉釀滿厚實的蝦膠，提升雙倍的鮮嫩蝦味，吃得精緻，令人讚賞不已。

材料

大蝦 1 隻（每隻約重 50-60 克）

蝦膠 40 克

生粉少許

鹽少許

葱花少許

蝦膠材料

海中蝦 1.2 公斤

肥肉粒 20 克

蛋白 1 個

鹽、糖各 4 克

生粉 12 克

獻汁

牛油 10 克

上湯 60 克

鹽少許

生粉水少許

蝦膠做法

- 參考本書第 24 頁製作蝦膠的做法。

做法

1. 大蝦去頭，去掉蝦殼，留蝦尾。加入鹽及生粉搓揉，用水沖淨，以毛巾壓乾水分。
2. 在蝦背輕切一刀，去腸，輕力拍扁，用刀輕力切斷蝦筋。
3. 大蝦抹上少許生粉，均勻地釀入蝦膠。
4. 燒熱油，放入大蝦以慢火煎蝦膠面，翻轉再煎蝦肉部分至七、八成熟，上碟。
5. 煮滾牛油及上湯，以鹽調味，下生粉水埋獻，淋於大蝦上，最後撒上蔥花即成。

在蝦肉輕切斷蝦筋，令蝦肉收縮後挺直不蜷曲，蝦滑與蝦肉黏合得更佳。

家鄉蒸龍躉腩

以順德大頭菜、陳皮及紅辣椒蒸龍躉腩，巧妙地辟除魚肉腥味，傳統而簡單的配搭，突顯龍躉的嫩滑和鮮味。

材料

龍躉腩 6 件（每件 38 克）
大頭菜 30 克（切絲）
紅辣椒 3 片
葱花、薑絲、冬菇絲、陳皮絲各少許

醃料

油、鹽、生粉各少許

料頭

蒜蓉、乾葱蓉、薑米各少許

做法

1. 龍躉腩洗淨，抹乾水分，拌入醃料待醃味均勻。
2. 用油爆香料頭，備用。
3. 龍躉腩放於碟上，撒入大頭菜絲、薑絲、冬菇絲、陳皮絲、紅辣椒及料頭。
4. 燒滾水，放入龍躉腩以大火蒸 4 分鐘，熄火，待 1 分鐘後，開蓋，潸葱花油及生抽，即可品嘗。

富哥秘訣

- 順德大頭菜經精心醃製而成，爽脆鹹香，適合配搭葷菜烹調，而且健脾開胃、潤肺止咳。
- 以大頭菜、陳皮、冬菇、薑絲等搭配龍躉腩，鹹香惹味的配料去掉魚腥味，更增添另一番農家風味。

福祿百寶鴨

富哥剛入行時已隨梁耀棠師傅炮製這道經典菜式，後來富哥將菜式改良，加添滋補及惹味食材釀入鴨身，經過長時間烹調，百寶鴨味道鹹香而不膩。

材料

全鴨 1 隻

薑 4 片

葱 2 條

八角 3 顆

上湯少許

餡料

鹹蛋黃 6 個（切粒）

蓮子 30 克

百合 30 克

洋薏米 30 克

栗子 80 克

銀杏 30 克

冬菇粒 10 克

金華火腿粒 10 克

糯米 30 克

上皮料

老抽少許

做法

1. 全鴨洗淨，去除內臟，整隻鴨去骨（保留鴨皮完整無損）。
2. 糯米蒸熟，與其他餡料拌勻炒香，釀入鴨腹內，用針線縫好，塗抹老抽上色（見圖左）。
3. 燒熱油，放入全鴨炸至定型，取出，瀝乾油分。
4. 全鴨放入大碟，加上薑、葱、八角及上湯，用大火蒸 1.5 小時，最後埋獻，伴菜遠享用。

- 百寶鴨的製作工序多，尤其去鴨骨時不能破壞鴨皮，整個烹調過程往往花好幾小時才能完成，是一道傳統手工菜。
- 富哥選用不肥膩的小穀鴨製作，肉質嫩滑，層次豐富。
- 添加了百合及栗子，健脾養胃、補腎、清心安神，提升滋補功效。
- 糯米與八寶料充分吸收鴨汁精華，味道融合，入口綿軟。

糖醋大蝦

一位用

糖醋大蝦是受歡迎的菜式，製作非常嚴謹，甜酸的糖醋汁恰到好處，富哥配上肉質爽滑的大蝦，令整道菜的口感提升不少，令人耳目一新。

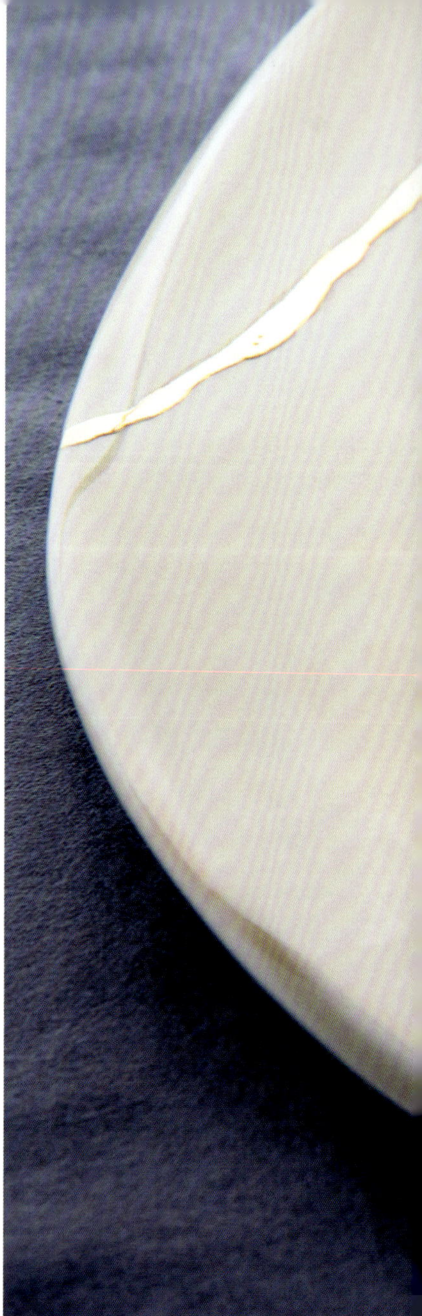

材料

大蝦 1 隻（每隻約重 85-100 克）

菠蘿 1 塊

雞蛋及生粉各少許

糖醋汁適量（製法參考第 70 頁）

做法

1. 大蝦去頭部沙囊，去掉中間蝦殼部位，用刀輕切開背，保留蝦頭及蝦尾，沖淨備用。
2. 大蝦蘸上蛋漿及生粉，放入熱油炸脆至熟，上碟。
3. 煮熱糖醋汁，淋於大蝦，最後以菠蘿伴碟享用。

將大蝦中間部位的蝦肉開邊，讓糖醋汁滲進蝦肉，保留原隻蝦頭及蝦尾，賣相大方得體。

大良蟹盒

順德大良的傳統名菜，肥肉切成薄片，兩塊相合，內藏蟹肉、蝦肉、豬肉及筍肉等等。肥肉切得愈薄，炸起的蟹盒更鬆化，絕對講究大廚的刀工技巧。

材料

肥豬肉 1 塊（約長 8cm、闊 8cm、厚 0.5mm）

紅色蟹黃少許

芫荽葉適量

餡料

豬柳肉粒 320 克

蝦粒 320 克

筍粒 400 克

蟹肉 240 克

韭黃粒 120 克

蛋白漿

蛋白 2 個

粟粉少許

水適量

做法

1. 肥豬肉用刀片成薄片。
2. 餡料拌勻；取一塊肥肉薄片，放入餡料 5 克、芫荽葉及紅色蟹黃，再蓋上一塊肥肉薄片，壓實。
3. 蛋白漿調勻，放入蟹盒蘸好，下熱油炸脆至鬆化，上碟。

富哥秘訣

- 必須掌握油溫及炸製時間，不能用油太猛，炸出來的蟹盒才能保持金黃色澤，而且香脆、無油膩感。
- 如果未能薄切成片，可將肥肉預早冷凍，然後以刨片機刨成薄片。

鮑汁蝦子柚皮

老香港的味道，是不少人的共同回憶。厚身的柚皮吸收鹹香的蝦子香氣，口感鬆化綿軟，入口甘香豐腴，是一道工藝逐漸失傳的香港特色傳統菜。

材料

柚皮 20 件

蝦子適量

調味料

蠔油及老抽各少許

燜料

鯪魚 800 克

排骨 400 克

瘦肉 200 克

蝦米 60 克

金華火腿骨少許

金華火腿皮少許

薑葱少許

做法

1. 柚皮刨去外皮，飛水，啤水後壓乾水分，浸泡一至兩天，以去掉苦澀味道，抹乾水分。
2. 燒熱油，放入柚皮炸至硬身，備用。
3. 鯪魚洗淨，去內臟；排骨及瘦肉洗淨，全部抹乾，備用。
4. 燒熱滾油，放入鯪魚、排骨及瘦肉炸香，盛起。
5. 薑葱、蝦米爆香，放入柚皮、燜料、調味料及水燜約 1.5 小時，至柚皮軟腍入味。
6. 柚皮上碟，灑上蝦子享用。

富哥秘訣

這道菜式的工序相當繁複，注意浸發柚皮的步驟，多次啤水及換水浸泡後，苦澀味基本上可去掉；否則燜煮後的柚皮帶苦味而不好吃。

脆皮燒乳鴿

外脆內嫩的燒乳鴿，最深得老饕青睞。富哥用心鑽研菜式，對於燒製乳鴿，他曾認真思考精進技巧，所以乳鴿外皮酥脆、肉汁豐富，令人食指大動。

材料

乳鴿 1 隻

麥芽糖及水各適量

伴吃

五香粉、喼汁各少許

做法

1. 乳鴿去內臟，洗淨，抹乾水分。
2. 麥芽糖水座於熱水調溶，備用。
3. 乳鴿用特製醬油醃味，塗抹麥芽糖水，吊掛風乾，備用。
4. 燒熱油，放入乳鴿先炸背部，再炸鴿胸位置，至色澤金黃勻稱，上碟，食用時灑上五香粉或喼汁，味道更突出。

富哥秘訣

- 由於鴿背翼位不容易熟透，所以先炸鴿背，再轉而油炸胸位，能夠保持乳鴿的外脆、肉汁鮮味，以免鴿胸太乾太硬。
- 還有一獨門秘笈，可將乳鴿上下倒過來油炸，這樣翼位張開來容易炸熟，能夠縮短接觸熱油的時間，保持鴿肉濕潤感。

鍋貼大明蝦

鬆脆的鍋貼大明蝦，多士炸得金黃，富哥配上鮮香的開邊中蝦，入口零油膩感，蝦肉與多士完美地配合，在口內拼發出鮮甜鬆化的口感。

材料

生中蝦 6-8 隻（每隻約 38 克）

白方包 4 片

芫荽葉 6-8 片

金華火腿蓉少許

蛋白漿

蛋白 2 個

粟粉少許

水適量

做法

1. 中蝦去殼，沖淨，開邊，挑去蝦腸。
2. 方包切件長方形，塗抹蛋白漿，放上開邊蝦黏好，蝦面貼上芫荽葉及火腿蓉。
3. 燒熱油，放入多士炸至香脆及蝦肉熟透，盛起，瀝乾油分，上碟享用。

富哥秘訣

- 蛋白漿是很好的黏合料，能夠緊緊地將明蝦黏貼麵包面，不易脫落。
- 炸油的油溫不要調得太高，否則多士容易炸至焦燶，影響賣相。

脆皮墨魚大腸

豬大腸釀滿黑松露墨魚膠，炸至香口乾身，外皮香脆，墨魚膠鮮味濃郁，肉質彈牙，在傳統的菜式上加添新意，讓食客品嘗到另一番食味體驗。

材料

豬大腸 1 條

薑葱、八角、香葉及鹽各少許

糖醋汁適量（製法參考第 70 頁）

墨魚膠材料

墨魚 250 克

黑松露碎 1 顆

肥肉粒 20 克

蛋白少許

鹽及生粉各少許

上皮水

白醋 200 克

麥芽糖 50 克

大紅浙醋 5 克

雙蒸酒 5 克

做法

1. 豬大腸徹底擦淨，洗淨。
2. 燒滾水，放入薑葱、八角、香葉及鹽，反轉豬大腸，內層向外，放入熱水煮 1 小時，去掉肥膏及油脂，熄火，加蓋待 15 分鐘，盛起。
3. 墨魚洗淨，隔乾水分，切成塊狀，加入配料打成墨魚膠，用手抓起墨魚蓉並撻回盤內，反複數次至呈膠狀。
4. 將墨魚膠釀入豬大腸，塗抹上皮水，吊掛風乾。
5. 豬大腸放入熱油炸脆至呈金黃色，瀝乾油分，切成斜片，伴糖醋汁上碟享用。

富哥秘訣

必須控制炸油的溫度及炸製時間，才能炸至外層香脆、內餡軟彈的墨魚膠豬大腸。

富哥

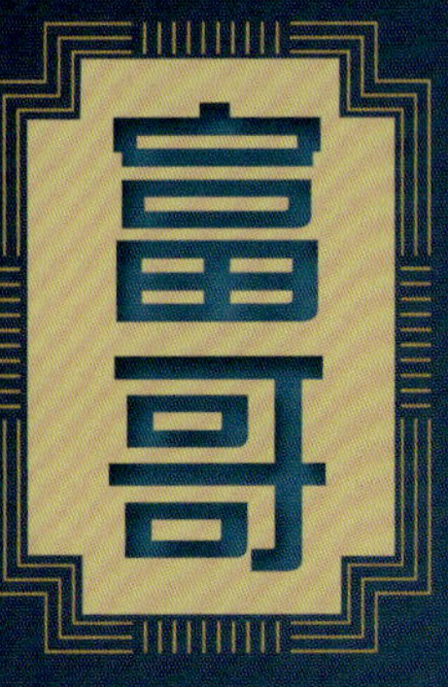

宴客佳饌

三五知己聚會，品酒之餘，以佳餚相伴，最是難得。
富哥的佳餚美饌，深得食客推崇——
焗釀響螺、蜜餞金蠔、香燒極品和牛……

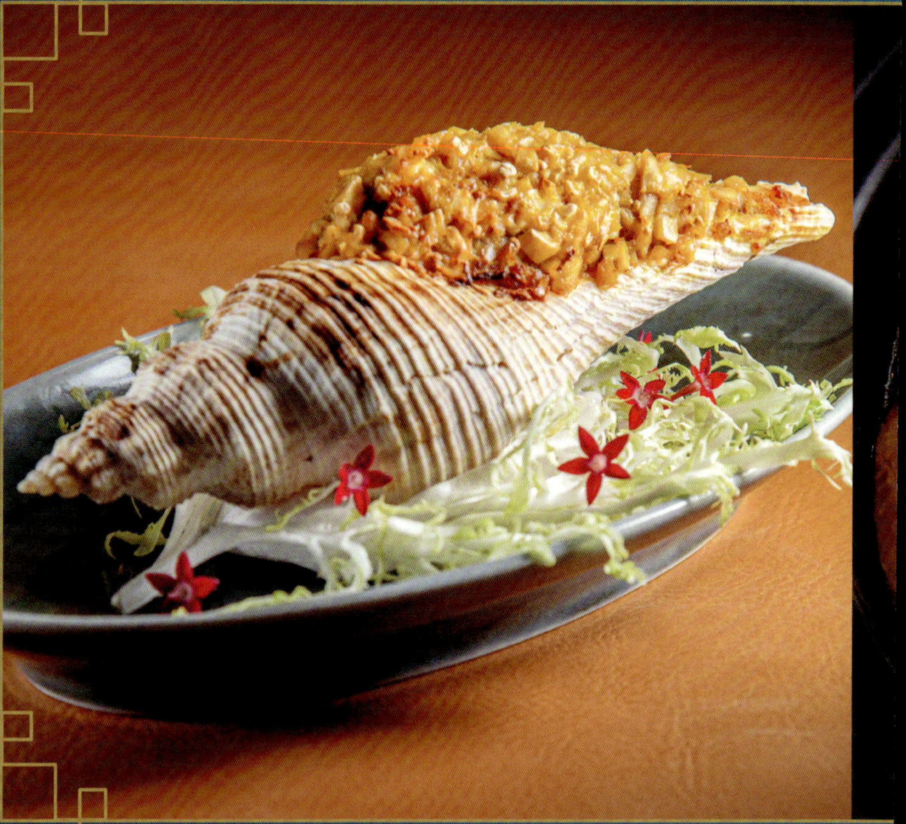

富哥曾擔任名人富豪家廚，深知食客的飲食品味，
練就他做菜的講究及認真，精益求精、求新求變，
不斷揣摩烹調竅門，盡顯具個人風格的品味菜。

蜜餞金蠔

選自南中國海無污染的生蠔，精製而成的金蠔，金黃飽滿、蠔香滿溢，煎香後外脆內軟，是富哥拿手的高級粵菜，宴請親友的指定菜式。

材料

生曬金蠔 6-8 隻

汁料

蜜糖 20 克

糖 10 克

生抽及老抽各少許

上湯 20 克

做法

1. 金蠔沖淨，抹乾水分，備用。
2. 燒熱水，放入金蠔隔水蒸熟，再放入油鑊煎香，盛起。
3. 煮滾汁料，放入金蠔拌勻，上碟享用。

富哥秘訣

- 生蠔益精補腎，富哥挑選南中國沿海的生蠔，在陽光之下原隻曬至半乾，蠔味鮮香撲鼻，齒頰留香。
- 伴上甜膩香濃的蜜糖汁，肥美的金蠔得以提升鮮味，是南中國海的美味水產美饌。

炒桂花瑤柱

這道古法粵菜，以雞蛋炒成金黃細粒桂花狀，伴以蟹肉、瑤柱等炒至爽口及清香，色澤鮮明引人，散發着獨特的家鄉風味，考驗大廚的功力。

材料

雞蛋 3 個

已浸發江瑤柱 40 克（約 1 兩）

銀芽 60 克

蟹肉 80 克

叉燒 20 克

葱花適量

韭黃少許（後下）

熟金華火腿絲少許

調味料

鹽、紹酒及麻油各少許

做法

1. 江瑤柱壓乾水分，放入半份用溫油炸香，盛起。
2. 銀芽用上湯煮至六成熟，盛起；雞蛋拌入鹽調味。
3. 燒紅鑊，放入蛋液、蟹肉、瑤柱、叉燒及葱花快炒成桂花粒狀，後下銀芽及韭黃炒勻，灑入鹽、紹酒及麻油增添香氣，上碟，最後綴以炸瑤柱及熟金華火腿絲裝飾。

雞蛋炒得幼細如桂花狀卻不乾身；芽菜需要保持爽口質感，口感層層遞進。

龍蝦蟹皇球

龍蝦要選用上好的貨色，配上富哥高超的手藝，龍蝦肉質爽滑，搭配蝦肉蟹皇球，三者的鮮味完美融合，加上龍蝦懾人的擺盤氣勢，食味及賣相同樣一絕。

材料

龍蝦 1 隻（750 克）

膏蟹 2 隻

雞湯適量

上湯及牛油各少許

青紅甜椒少許（切粒）

麵包糠少許

鹽少許

蝦膠材料

中蝦 600 克

肥肉粒 20 克

蛋白 1 個

鹽、糖及生粉各適量

蝦膠做法

- 參考本書第 24 頁製作蝦膠的做法。

做法

1. 膏蟹拆出蟹膏，與雞湯煮熟，用手搓成粒狀。
2. 蝦膠釀入蟹膏，包好搓圓，滾上麵包糠，放入熱油內炸脆，瀝乾油分。
3. 龍蝦去頭殼，蒸熟，保留作為裝飾。
4. 龍蝦肉去殼，切件，拉油五成熟，備用。
5. 燒熱上湯及牛油，放入青紅椒粒炒煮，加入龍蝦肉拌炒，灑入鹽調味，埋獻，上碟。
6. 蝦膠蟹皇球伴龍蝦球旁，以龍蝦頭裝飾即成。

富哥秘訣

- **打蝦膠最好選用蝦公，肉質爽口。**
- **龍蝦肉拉油至五成熟即可，稍後需回鑊與配料炒煮，以免肉質太老。**

香燒極品和牛

和牛放入煎鑊嗞嗞作響，繼而散發香濃的牛肉香氣，鮮、嫩、香，入口即溶，品嘗到滿口肉汁的鮮甜滋味，令人再三回味。

材料

和牛肋肉 200 克

燒汁 20 克

沙律菜少許

醃料（醃牛肉 6 公斤計算）

雜菜水適量

鹽及糖各少許

雞蛋 6 個

炒胡椒粒少許

做法

1. 和牛肋肉以醃料拌勻，醃一夜，備用。
2. 燒熱煎鑊，放入和牛肉煎至七成熟，切片，上碟。
3. 煮熱燒汁，淋於和牛肉上，伴以沙律菜享用。

富哥秘訣

- 要控制和牛肋肉的熟度，煎至七成熟即可，切出來的牛肉粉粉嫩嫩，入口嫩滑、肉汁豐腴。
- 醃料的雜菜水，以洋葱、紅蘿蔔及西芹切塊，放入油鑊爆香，下香葉及八角炒香，濳酒，加水煮成雜菜水。

焗釀響螺

一位用

響螺肉混和雞肉及燒雞肝，完美地互相搭配，釀入響螺殼焗製，飽滿的餡料令食客按捺不住品嘗，鮮、香、辛，是一場無與倫比的品味之旅。

材料

響螺 1 隻（約 240 克）

熟雞粒、燒雞肝粒各適量

蘑菇粒及草菇粒適量

牛油及忌廉奶各少許

咖喱粉 1 茶匙

上湯 40 克

蛋黃 1 個

響螺殼 1 個

料頭

蒜蓉、乾葱蓉、薑米及

洋葱粒各少許

做法

1. 響螺焓熟，拆肉，切碎備用。
2. 燒熱油，放入料頭及咖喱粉爆香，加入上湯煮熱，下響螺肉、熟雞粒、燒雞肝粒、蘑菇粒及草菇粒炒香，加入牛油及忌廉奶拌勻至略乾，餡料盛起。
3. 將響螺肉放於響螺殼中間位置，填入響螺洋葱餡料。
4. 餡料面輕掃蛋黃漿，放入焗爐以 220℃焗 12 分鐘至表面金黃，即成。

響螺肉質感彈牙，肉質鮮美，與雞肉、燒雞肝及菇菌完美匹配，啖啖響螺肉海鮮之味，富有嚼勁。

雙冬枝竹羊腩煲

古法炮製的羊腩煲，搭配冬菇及冬筍燜煮，羊肉嫩滑濃香，羊脂肥肉爽彈有致，油香豐盈，枝竹吸收羊肉的羶味精華，大廚的手藝應記一功。

材料

羊腩 1.8 公斤

冬菇 100 克

冬筍粒 240 克

枝竹 240 克

陳皮 2 個（切絲）

薑葱各少許

蒜子 100 克

燜羊腩醬 2 湯匙

紹酒少許

做法

1. 冬菇浸軟，剪去硬蒂；冬筍去掉外皮，洗淨，切塊；陳皮浸軟，刮去內瓤。
2. 羊腩沖淨，原件飛水，瀝乾水分，塗抹老抽上色。
3. 燒熱油，放入羊腩炸香，盛起，再用油鍋煎香去掉羶味，沖淨備用。
4. 下油燒熱，放入薑、葱及蒜子爆香，下羊腩、冬菇及冬筍拌勻，灒酒，拌入燜羊腩醬炒勻，加入熱水以小火燜煮約 1.5 小時至軟腍。
5. 枝竹燜好，放入瓦鍋底；羊腩切件，放於瓦鍋，羊腩汁埋獻，淋上即可品嘗。

富哥秘訣

- 羊腩需要長時間燜煮，炒煮配料後，建議竹笪放鍋底，以免羊腩黏着燒焦，保持足夠醬汁燜煮即可。
- 原件羊腩燜煮，保留羊腩肉汁，最後切塊後埋獻，嚐到滿口羊肉香氣。

西施大蝦球

一位用

雪白剔透的大蝦球，美人如玉似西施，讓食客細意觀賞，滿足視覺享受之餘，再品嘗爽口彈牙的大蝦球，視覺及味覺同樣精彩滿分。

材料

大蝦 1 隻（每隻約 100-150 克）

雲南火腿 1 片（切絲）

醃料

鹼水少許

梳打粉少許

蛋白料

蛋白 1 份

鮮奶 1 份

上湯 1 份

做法

1. 大蝦沖淨，去腸，用刀薄片去皮，剔成花狀。
2. 大蝦以鹼水及梳打粉拌勻，醃 2 小時，放於水喉下沖水至蝦身爽滑。
3. 大蝦飛水，瀝乾水分，放入熱油拉油，待蝦肉轉成雪白色。
4. 將蛋白料拌勻，過濾，隔水蒸成蛋白底。
5. 放上大蝦球，淋上湯獻，以火腿絲裝飾品嘗。

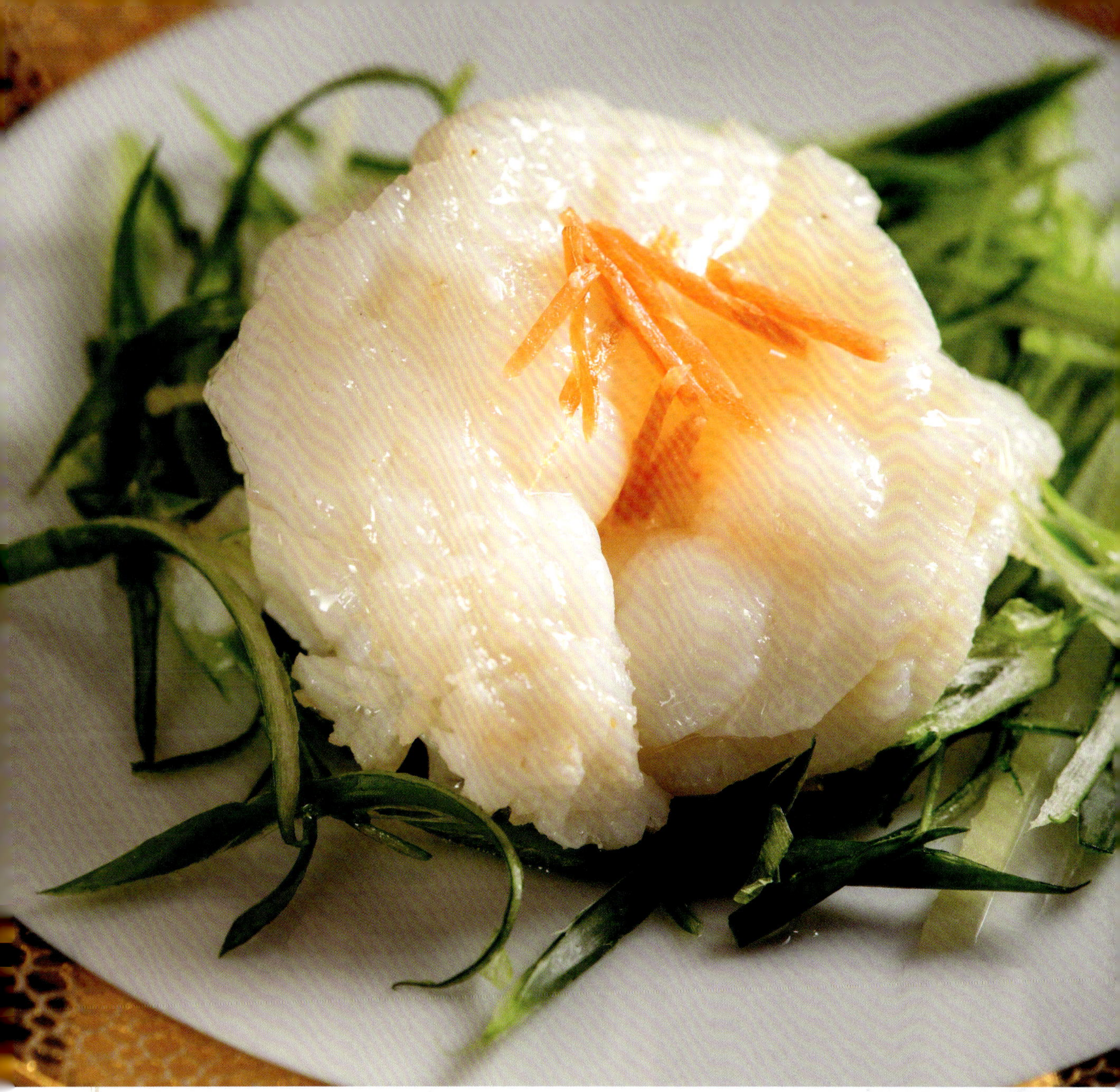

富哥秘訣

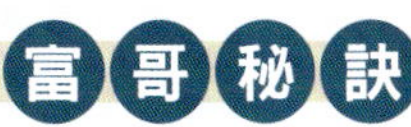

- 西施大蝦球以雪白見稱，要注意必須徹底去掉紅色蝦皮，才能呈現雪白的蝦球。
- 以鹼水及梳打粉拌醃，蝦肉爽滑有彈性。

頭抽王焗蠔

頭抽是黃豆經發酵後，第一趟提取出來的醬油，是醬油的精華所在。生蠔裹上薄粉炸脆，以濃郁味鮮的頭抽拌和，醬油與生蠔兩者的鮮味拼發而出。

材料

新鮮大鮮蠔 8 隻

頭抽 15 克

葱段少許

紹酒少許

蛋白漿

蛋白 2 個

粟粉少許

水適量

做法

1. 鮮蠔用生粉擦淨，用水沖淨，放入滾水浸 4 分鐘，取出，瀝乾水分。
2. 鮮蠔蘸上蛋白漿，放入滾油炸至金黃色，盛起。
3. 燒熱油，下葱段爆香，放入炸鮮蠔及頭抽拌勻，灒酒，上碟品嘗享用。

大鮮蠔裹上薄薄的蛋白漿即可，嚐到薄脆的外皮。

土魷剁黑毛豬肉餅

土魷蒸肉餅是尋常百姓家的家常菜，富哥提升飲食質素，選取黑毛豬頸骨下的爽肉（梅頭第一刀），肥瘦勻稱，肉質軟腍，嚐到肉嫩多汁的肉餅，口感彈牙爽口。

材料

梅頭豬肉（第一刀）200 克

土魷 25 克

冬菇粒 25 克

醃料

蛋白 1 個

鹽、糖各 2 克

生粉 4 克

生抽、老抽、油及水各少許

做法

1. 梅頭豬肉洗淨，切粒，用刀剁成肉碎。
2. 土魷撕去外衣，用水浸軟，切成幼粒，備用。
3. 肉碎、土魷粒、冬菇粒及醃料全部放進大碗內，以順時針方向拌勻至略帶黏稠，放於蒸碟。
4. 燒滾水，肉餅隔水以大火蒸 6 分鐘熟透，上桌享用。

富哥秘訣

每頭豬只有 450 克第一刀梅頭爽肉，十分矜貴。手工剁成的薄肉餅，肉味濃郁，爽口帶嚼勁，肉汁豐富，為傳統的蒸肉餅添上新意。

金錢海鮮盒

富哥將傳統菜式融和，用麵包裹好餡料，油炸而成金黃色的金錢圓塊，賣相精緻高雅，而且嚐到龍蝦、韭菜及馬蹄等餡料，令食客印象難忘。

材料

白方包 3-4 片
薯蓉少許
冬菇絲 12-16 小片
芫荽葉 3-4 片
金華火腿蓉少許
蛋白 1 個

餡料

龍蝦粒 28 克
馬蹄粒 28 克
韭菜粒 28 克

調味料

鹽、生粉、蛋白各少許

做法

1. 每塊方包用圓形模具裁切成圓形；龍蝦餡料拌勻，備用。
2. 薯蓉與調味料拌勻至幼滑。
3. 圓形方包內放入餡料 20 克，蓋上薯蓉抹好，加上冬菇絲、芫荽葉及金華火腿蓉裝飾。
4. 海鮮盒面以蛋白均勻地塗抹，放入熱油炸至金黃色，上碟品嘗。

富哥秘訣

- **餡料可隨個人喜好調配，蟹肉、蝦肉、沙葛或韭黃皆可，做出自己心目中的金錢海鮮盒。**
- **包入餡料後，用薯蓉蓋好餡料，保存海鮮的味道，也免油炸時餡料四溢等情況發生。**

豉油王大蝦

惹味的豉油王大蝦，是與老友把酒言歡的指定菜式。以薑、葱、蒜等料頭炒香的大蝦，味道沒法抗拒，富哥灑上頭抽王抄拌，加上大蝦上碟的非凡氣派，絕對是大宴親朋之選。

材料

大蝦 600 克（每隻 75-100 克）

頭抽王適量

薑米、乾葱、蒜蓉及葱花各少許

做法

1. 大蝦保留外殼，在背部剔一刀開背，去掉蝦腸，沖淨。
2. 燒熱油下薑米、乾葱、蒜蓉及葱花爆香，放入大蝦，灑入頭抽王拌抄勻，先炸大蝦至八成熟，炒勻料頭即可上碟。

富哥秘訣

- **在蝦背稍剔一刀開背，令料頭及頭抽精華滲進蝦肉。**
- **又或可將蝦頭及蝦身分開，先炸蝦頭，再放入蝦身炸至八成熟，最後一併以頭抽王及料頭拌抄，惹味滿分。**

竹笙燴上素

榆耳補益和中、固腎氣；黃耳稱為素燕窩，滋陰潤肺、補脾胃，這道素菜配搭爽滑的竹笙及菇香飄溢的菇菌，是宴客葷菜以外的不二之選。

材料

榆耳、黃耳、冬菇、草菇、蘑菇、冬筍、竹笙及髮菜共 300 克

銀芽少許

菜遠 4 棵

上湯適量

做法

1. 榆耳、黃耳、冬菇及竹笙分別浸泡清水。
2. 髮菜用水浸 20 分鐘，拌入油輕揉，用水沖洗 3-4 次。
3. 榆耳、黃耳、冬菇、草菇、蘑菇、冬筍及竹笙飛水 2 次，盛起，瀝乾水分。
4. 燒熱油，放入所有菇菌及冬筍爆香，用上湯浸過煨透入味，盛起。
5. 菜遠及銀芽飛水，盛起。
6. 所有菇菌、冬筍及髮菜用上湯燜入味，上碟，伴以菜遠及銀芽享用。

富哥秘訣

- 榆耳及黃耳質感有點硬，必須先用上湯煨煮入味至軟身。
- 茹素人士宜用素湯代替上湯。

紅燒蘿蔔牛腩

富哥在會所工作時，紅燒蘿蔔牛腩是他的招牌菜式，以羊鴨醬、柱侯醬及麵豉醬調成濃濃的醬汁，燜煮的牛腩香濃惹味，是累積多年來的經驗與手藝，平凡中蘊含工夫。

材料

牛坑腩 1 件（約 1.2 公斤）

牛筋 600 克

蘿蔔適量

八角 3 顆

薑片及葱段各少許

上湯適量

醬料

羊鴨醬 40 克

柱侯醬 40 克

麵豉醬 10 克

冰糖 15 克

做法

1. 牛筋沖淨，飛水，處理乾淨，隔水蒸 1 時 15 分鐘至軟腍。
2. 牛坑腩沖淨，飛水，隔乾水分，用油煎香，盛起。
3. 燒熱油，放入薑片及葱段爆香，下羊鴨醬、柱侯醬及麵豉醬炒香，放入牛坑腩、牛筋、上湯及冰糖燜約 1 小時 15 分鐘。
4. 蘿蔔燜約 15 分鐘至入味，鋪於鍋底；待牛坑腩及牛筋軟腍，切件，放入鍋內，埋獻即可享用。

富哥秘訣

- 羊鴨醬是混和薑、蒜等辛香料的麵醬，與豉味香濃的柱侯醬，以及黃豆發酵而成的麵豉醬，三者融和，互相調配，令羊腩的味道豐富多采。
- 醬料必須用薑葱料頭爆透，才能爆發香味。

大明蝦沙律

在炮製傳統粵菜佳餚之餘，富哥將西式美食與中菜融匯入饌，這道大明蝦沙律是其中好例子，清甜的中蝦配襯西芹、菠蘿及蘋果等蔬果，清爽開胃。

材料

大中蝦 900 克

西生菜 100 克

西芹 30 克

蘋果 2 個

菠蘿 1 個

芒果 2 個

士多啤梨 6 顆

櫻桃 6 顆

沙律汁

沙律醬 240 克

熟蛋黃 2 個

檸檬 1 個（榨汁）

鹽及煉奶各少許

做法

1. 將沙律汁材料拌勻，成為香滑沙律醬。
2. 所有生果去皮、去芯，切粒；西芹撕去硬筋，西生菜洗淨，切塊。
3. 大中蝦焓熟，去殼，挑去腸，開邊。
4. 將蔬果鋪於碟上，淋上沙律醬汁拌勻，放上開邊熟蝦，即可享用。

富哥秘訣

- 這款沙律汁味道蘊含蛋黃香、檸檬酸香及煉奶甜膩，配搭鮮果品嘗，清甜可口。
- 鮮果及大中蝦層層排列，顏色豐盛，引人食慾。

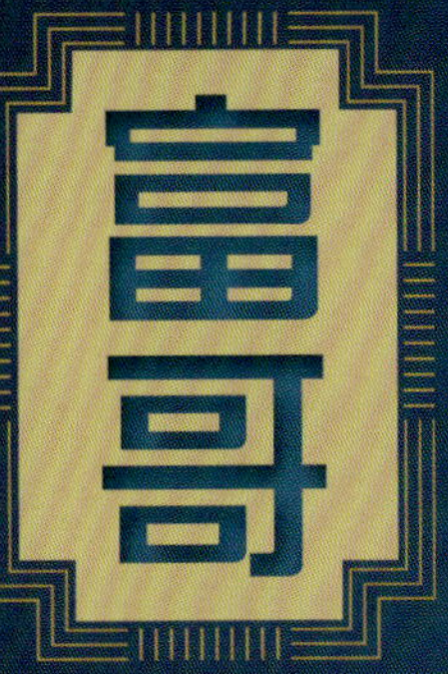

招牌飯麵

炒飯炒麵，最講究鑊氣，
動作有序、不急躁，還要掌握火候與食材的平衡點。

富哥招牌炒飯、蟹肉炒米粉，堪稱富哥的招牌主食，
在五分鐘的炒煮過程，是一場以廚房為舞台，
撒入材料及調味料、舞動鍋子的精彩表演，
最後快炒一刻，完美落幕。

富哥招牌炒飯

一九八零年，富哥跟隨粵菜大師梁耀棠師傅學習廚藝，梁師傅做菜特別認真，炒飯用的米，用上湯蒸熟，炒出來的米飯與眾不同，顆粒分明，特別香口，堪稱一絕。

材料

小黃蝦 140 克

金華火腿碎 20 克

雞蛋 2 個（拂勻）

葱花 76 克

蒸米飯料

白米 200 克

上湯 200 克

生油少許

做法

1. 白米、生油及上湯拌和，隔水蒸熟後拌鬆，待涼，取米飯約 320 克。
2. 小黃蝦去殼，洗淨，飛水，用油略炒，盛起；葱花吸乾水分。
3. 燒熱油，放入蛋液炒至八成熟，拌入米飯壓平受熱平均，灑入金華火腿碎，以大火炒至米飯乾透，放入鮮蝦，灑上葱花炒熱即可。

富哥秘訣

- 富哥將鑊燒得極熱，炒出來的米飯粒粒分明，油分不多，米粒散發淡淡上湯的香氣，還要不停手炒，鑊氣十足，否則飯粒容易燒焦。
- 白飯必須用上湯新鮮即蒸，不能用隔夜飯，否則太乾太硬。
- 小黃蝦比一般鮮蝦嫩滑，而且鮮甜美味。

牛柳絲炒麵

煎得香口卜脆的麵條，以牛柳絲及青紅甜椒伴吃，咬入口香脆的聲響，味蕾嚐到的是濃香及爽脆拼湊的結合，一塊接着一塊吃不停。

材料

牛柳絲 120 克

銀芽 100 克

鹹酸菜、青紅甜椒共 100 克

炒麵餅 1 個

調味料

上湯 300 克

老抽少許

鹽少許

生粉水適量

做法

1. 鹹酸菜及青紅甜椒切絲，備用。
2. 炒麵餅放入熱水燙開，用水沖透，去掉鹼水味，晾乾備用。
3. 燒熱油，放入麵餅煎至兩面微黃及香脆，剪成 6-8 份，排放碟上。
4. 燒熱油，放入牛柳絲、鹹酸菜及甜椒炒勻，加入上湯、老抽及鹽調味，以生粉水埋獻，放於脆麵餅上，趁熱伴吃。

富哥秘訣

- 麵餅煎得香脆是關鍵所在，要加入適量食油煎香，並略壓成扁平狀。
- 炒麵餅可在粉麵店購買，每個約 3 至 4 兩。

蟹肉炒米粉

富哥炒米粉非常出色，米粉炒得乾身，彈牙爽口，條條分明，而且帶香口的鑊氣，讓食客感受到蟹肉與米粉融和的絕佳香氣。

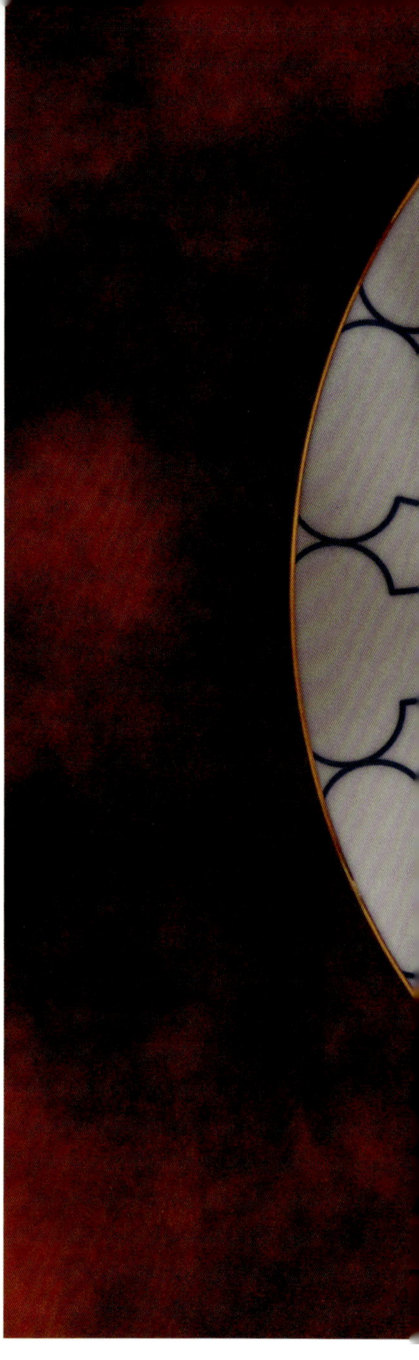

材料

米粉 320 克

熟蟹肉 80 克

銀芽 140 克

洋葱絲 40 克

雞蛋 2 個

上湯適量

葱花 30 克

調味料

老抽及鹽各少許

做法

1. 米粉放入凍水，慢火加熱，水燒滾後打散米粉，盛起，混和幾滴熟油，用乾淨布包好。
2. 洋葱絲炒熟，加入銀芽和上湯炒至六成熟，盛起。
3. 燒熱油鍋，下雞蛋炒至七成熟，放入米粉煎香，灑入鹽及老抽炒勻，最後加入熟蟹肉、洋葱絲及銀芽，最後灑入葱花炒香上碟。

富哥秘訣

- 富哥選用廣東河源出產的米粉，炒出來的米粉特別爽滑而不爛。
- 打米粉後拌和油分，用布包好，目的是用熱力令米粉變軟。
- 米粉上碟前，用沾上油的鑊鏟快炒 5 秒，能散發香氣撲鼻的米粉香。

松露和牛河

由乾炒牛河變奏而成，以澳洲和牛搭配黑松露拌炒河粉，保持粵菜傳統之餘，創意十足，為食客帶來味道新衝擊，是絕佳的新嘗試。

材料

澳洲和牛肩肉 160 克

河粉 380 克

青、紅甜椒 25 克（切絲）

銀芽 38 克

葱絲少許

黑松露碎適量

調味料

生抽、老抽各少許

做法

1. 和牛肩肉去筋，切薄片，排於小碗內疊好。
2. 燒熱油，放入銀芽炒至剛好，盛起。
3. 繼續下甜椒絲及河粉一起炒香，灑入調味料拌炒均勻，再加入銀芽炒勻，轉放排好和牛的小碗內。
4. 將小碗反轉蓋在大碟上，拿走小碗，以火槍燒熟牛肉至七成熟上桌，最後以黑松露碎及葱絲裝飾即成。

富哥秘訣

- 和牛以火槍燒至七成熟即可，能品嘗嫩滑質感的和牛風味。
- 銀芽炒至挺身，剛熟而不能釋出水分或過熟。

鮮蝦蟹肉飄香荷葉飯

炎炎盛夏，是荷葉的當造季節，用荷葉包裹的米飯，特別荷香撲鼻，散發陣陣幽香。鮮蝦、蟹肉、江瑤柱及火鴨，為荷葉飯配上完美的口感。

材料

鮮蟹肉 28 克

小黃蝦 120 克

江瑤柱 50 克

叉燒粒、火鴨粒各 40 克

筍粒 60 克

冬菇粒 38 克

荷葉 1 塊

米飯 380 克

紹酒少許

雞油少許

蒸米飯料

白米 200 克

上湯 200 克

生油少許

做法

1. 蒸米飯料拌好，蒸熟米飯，取米飯約 320 克，備用。
2. 小黃蝦去殼，洗淨；江瑤柱浸透，撕成絲。
3. 荷葉洗淨，放入熱水略燙，瀝乾水分。
4. 蟹肉、小黃蝦、江瑤柱、冬菇粒及筍粒，用上湯煨好入味。
5. 燒熱油，放入叉燒粒、火鴨粒及上述材料爆香，放入米飯炒好，灒紹酒及雞油，拌勻。。
6. 鋪好荷葉，底部放上蟹肉及江瑤柱，鋪上米飯包好，用大火蒸 20 分鐘即可。

富哥秘訣

- 米粒及上湯的比例是一比一，如一斤米則用上湯一斤蒸熟。
- 荷葉先放入熱水燙一下，才能逼出荷葉香氣。

香煎蘿蔔糕

富哥的蘿蔔糕，獨特地以老雞、金華火腿熬成上湯，用以調稀粘米粉及煮蘿蔔絲，故蘿蔔糕的味道格外清甜美味，難怪成為不少熟客的心頭好。

材料（製成 600 克蘿蔔糕）

蘿蔔 4.2 公斤（連皮計）

蝦米 120 克

臘腸 4 孖

膶腸 2 孖

江瑤柱少許

炒香白芝麻少許

上湯 1.6 公斤

粉漿料

粘米粉 600 克

粟粉 160 克

澄麵 160 克

上湯 800 克

調味料

片糖 35 克

鹽 40 克

做法

1. 蘿蔔洗淨，刨去皮，切成粗絲。
2. 臘腸及膶腸切粒；蝦米切蓉；江瑤柱浸軟，撕成絲。
3. 粘米粉、粟粉、澄麵及上湯調成粉漿，備用。
4. 燒熱油鑊，放入臘腸、膶腸及蝦米炒香，下蘿蔔絲、片糖、鹽及上湯煮滾，拌入粘米粉漿調勻。
5. 蘿蔔粉漿盛於容器內，以大火蒸 1 小時，最後灑入江瑤柱絲及芝麻，再蒸一會，煎香即可品嘗。

富哥秘訣

- 以老雞、金華火腿及瘦肉熬成上湯，代替水調成粘米粉漿；另用上湯炒煮蘿蔔絲，吸收上湯的精華。
- 挑選重身的蘿蔔，水分足夠，清甜可口。
- 蘿蔔必須用刀手切，粗幼有致，既可煮成蓉，又能吃到蘿蔔粗絲的口感，層次分明。

Contents

MASTER FU'S BANQUET FAVOURITES

MASTER FU'S SIGNATURE RICE AND NOODLES

Chicken Wing Stuffed with Bird's Nest

*Chinese version refer to p.14/makes 1 serving

INGREDIENTS:

1 fresh chicken winglet with tip
20 g bird's nest (rehydrated)
premium stock
caltrop starch
maltose (diluted with water)

MARINADE:

salt to taste

METHOD:

1. Cut a winglet from a dressed chicken. Keep the wing tip. Debone the wing flat. Set aside.
2. Simmer rehydrated bird's nest in premium stock until flavourful. Drain. Cook in a glaze until thickened. Set aside.
3. Stuff the deboned wing flat with 20 g of bird's nest. Secure the seam with a toothpick. Add salt and rub evenly. Leave it for 3 hours.
4. Boil a small pot of water. Put in the stuffed chicken wing and cook until the skin tightens. Brush some diluted maltose on the skin. Hang the wing at a windy spot to air dry for at least 2 hours until the skin is dry to touch.
5. Heat oil in a deep-fryer up to 120°C. Put in the wing and soak it in the oil until cooked through. Then turn up the heat to 160°C. Deep-fry the wing until crispy and golden. Drain and wipe dry with paper towel. Arrange on serving plate and garnish. Serve.

MASTER FU'S TIPS:

For this recipe, shop for bird's nest from swiftlets. Soak it in water overnight and simmer it in premium stock for the bird's nest to pick up the umami from the stock. It works very well with the flavoursome chicken wing.

Master Fu's Signature Dried Abalone

*Chinese version refer to p.16

INGREDIENTS:

3 kg dried abalones
2 fresh chicken (about 3.6 kg)
1.2 kg lean pork
1.2 kg pork belly ribs
600 g chicken feet
pork skin
10 dried scallops
spring onion

SEASONING:

20 g rock sugar
dark soy sauce

METHOD:

1. Soak the dried abalones in water until soft, for about 1 week. (The soaking time depends on their sizes.)
2. Steam the abalones over boiling water until tender, for about 8 hours. (Again, the steaming time depends on their sizes.) Drain and set aside.
3. Rinse the chicken and chop into pieces. Rinse the lean pork and pork belly ribs. Cut into pieces. Rinse the chicken feet and pork skin. Soak dried scallops in water until soft.
4. Heat oil in a wok. Stir-fry spring onion until fragrant. Put in chicken, lean pork, pork belly ribs, chicken feet and pork skin. Fry until fragrant. Set aside.
5. Put a bamboo mat on the bottom of a clay pot or ceramic pot. Arrange the fried ingredients from step 4 over the bamboo mat. Then add dried scallops, abalones and seasoning. Bring to the boil and turn to low heat. Simmer for 3 hours until the abalones are tender. Arrange abalones on a serving plate. Discard all solid ingredients and strain the sauce in the pot. Drizzle the abalones with the strained sauce. Add blanched vegetables and garnishes on the side. Serve.

MASTER FU'S TIPS:

- For this recipe, Master Fu uses Yoshihama dried abalones from Japan for their deep seafood flavour, tender texture, and centres in golden amber colour after cooked.
- To elevate the umami of abalones, soak them in premium stock before simmering them with other ingredients. They will soak up the flavours from the stock them way, with their umami boosted in major ways prior to the cooking process.

Deep-fried Pipa Tofu

*Chinese version refer to p.18

INGREDIENTS:

6 pieces tofu
200 g minced dace
5 egg whites
30 g Jinhua ham
38 g shallot
38 g coriander
20 g shredded dried tangerine peel
wheat starch

SEASONING:

10 g salt
10 g sugar
60 g caltrop starch

METHOD:

1. Finely dice the tofu. Drain well and leave it for 30 minutes.
2. Finely chop the Jinhua ham. Finely chop shallot and coriander. Set aside.
3. In a mixing bowl, put in minced dace, egg whites, Jinhua ham, shallot, coriander, dried tangerine peel and seasoning. Mix well. Stir in the diced tofu.
4. Divide the tofu mixture among ceramic spoons. Transfer into a steamer and steam for about 6 minutes until set. Leave them to cool.
5. Turn the tofu pieces out from the ceramic spoons. Coat them in wheat starch. Deep-fry in hot oil until golden. Dish up and serve.

MASTER FU'S TIPS:

Instead of minced dace, you may also use minced shrimp and add finely chopped spring onion to the tofu mixture. That would imbue this dish with different flavours.

Stir-fried Pork's Tripe-tip with Black Beans and Chili Sauce

*Chinese version refer to p.20

INGREDIENTS:

4 to 6 pork tripe (you'd need about 160 g of tripe-tips for this recipe)
160 g tricolour bell peppers and onion
2 tsp fermented black beans
diced ginger
grated garlic
chopped shallot

MARINADE:

1 swimmer crab (that has recently molted; squeeze to extract its juices)
4 g sugar
12 g caltrop starch

SEASONING:

salt
sugar
dark soy sauce

METHOD:

1. Rub coarse salt on both the inside and the outside of the pork tripe. Rinse well. Use only the fleshiest tips for this recipe. Remove any fat and membrane that adhere to the tripe. Slice thinly.
2. Add marinade to the pork tripe. Mix well. Leave it for 1 hour.
3. Deseed the bell peppers. Rinse and cut into chunks. Peel the onion. Rinse and cut into chunks.
4. Heat oil in a wok. Stir-fry ginger, garlic, shallot and fermented black beans until fragrant. Add bell peppers, onion and pork tripe tips. Toss quickly over medium heat. Drizzle with seasoning. Toss again. Serve.

MASTER FU'S TIPS:

- The tip only makes up one-sixth of a whole pork tripe. Thus, in order to have enough tripe tips to make a full-portion dish, you'd need at least 4 pork tripe. That's why it's considered a luxurious Cantonese recipe.
- You also need to control the heat precisely. To retain its tender but toothsome texture, stir-fry the pork tripe over high heat.

Stuffed Crab Claw with Shrimp Mousse

*Chinese version refer to p.22/makes 1 serving

INGREDIENTS:

280 g claw of a swimmer crab
60 g minced shrimp filling
2 slices sandwich bread

MINCED SHRIMP FILLING:

600 g medium shrimps
20 g diced fatty pork
1 egg white
salt
sugar
caltrop starch

METHOD:

MINCED SHRIMP FILLING:

1. Shell and devein the shrimps. Rub coarse salt on the shrimps. Rinse and press dry using a towel.
2. Put the shrimps on a chopping board. Spread the shrimps using the flat side of a cleaver. Then coarsely chop them using the back of a cleaver.
3. Put the chopped shrimps into a mixing bowl. Add egg white and stir in one direction until sticky. Add diced fatty pork, salt, sugar and caltrop starch. Mix well. Lift the chopped shrimp mixture from the bowl and slab it back into the bowl repeatedly until sticky. This is the minced shrimp filling.

ASSEMBLY AND DEEP-FRYING:

1. Rinse the crab claw. Remove any broken shell. Wipe dry.
2. Dice the sandwich bread. Leave it to dry.
3. Wrap the crab claw with the minced shrimp filling. Coat it with breadcrumbs from step 2.
4. Heat oil in a wok. Deep-fry the stuffed crab claw over medium heat until the crab claw floats and is cooked through. Keep on frying until golden. Dish up and garnish. Serve.

MASTER FU'S TIPS:

- By wrapping the crab claw in minced shrimp filling, this dish presents two different textures and umami at the same time.
- Instead of powder like breadcrumbs, I prefer diced bread for this recipe because of its crispy texture after fried.

Stuffed Sea Cucumber with Cream Sauce

*Chinese version refer to p.26/makes 1 serving

INGREDIENTS:

1 Hokkaido spiny sea cucumber (about 60 to 66 g before rehydration)
ginger (sliced)
spring onion (cut into short lengths)
garlic cloves
10 g crabmeat
20 g onion (shredded)
grated garlic
whipping cream
premium stock
butter
roux

EGG WHITE BATTER: (MIXED WELL)

2 egg whites
40 g corn starch
20 g caltrop starch

CHAMPAGNE SAUCE:

whipping cream
premium stock
butter
champagne (or any dry sparkling white wine)

GARNISHES:

diced green and red bell pepper
deep-fried rice grains

METHOD:

1. Soak the dried sea cucumber in water until soft. Drain and set aside.
2. Add ginger, spring onion and garlic cloves to the premium stock. Put in the sea cucumber and simmer until it picks up the flavours. Drain.
3. Heat oil in a wok. Stir-fry grated garlic, shredded onion and crabmeat briefly. Add whipping cream, premium stock and butter. Add roux and cook until it thickens and the crabmeat turns slightly stiffer. This is the filling.
4. Stuff the sea cucumber with the filling from step 3. Dip it into the egg white batter.
5. Deep-fry the sea cucumber in hot oil until crispy and golden. Save on a serving plate.
6. To make the champagne sauce, cook all ingredients and add caltrop starch slurry until thick. Add diced green and red bell pepper. Drizzle on the serving plate around the sea cucumber. Sprinkle with deep-fried rice grains. Serve.

MASTER FU'S TIPS:

- When making the filling, the crabmeat should be cooked until slightly stiff. Otherwise, the filling will be too mushy after being stuffed into the sea cucumber.
- To make the roux, melt butter in a pan and stir in plain flour until it turns into a thick paste. It is a stable thickener used in many Western dishes.

Baked Free Range Chicken with Shaoxing Wine

*Chinese version refer to p.30

INGREDIENTS:

1 freshly slaughtered chicken (about 1.8 kg)
200 g Shaoxing wine
spring onion (cut into short lengths)

MARINADE:

16.8 g salt

METHOD:

1. Rinse the chicken and remove all innards. Wipe dry. Rub salt all over and leave it for 4 hours.
2. Line the bottom of an electric rice cooker with spring onion. Put in the chicken with the breast side up. Drizzle with Shaoxing wine. Turn on the rice cooker to start a regular rice cooking cycle. Cook for 40 minutes until the chicken is cooked through. Slice and serve.

MASTER FU'S TIPS:

- When marinating the chicken, those parts with thicker flesh require more salt. Generally speaking, for a dressed chicken about 1.8 kg, use 5.6 g of salt for every 600 g of chicken.
- Lining the bottom of an electric rice cooker or claypot with spring onion and putting in the chicken with the breast side up help retain the juices. The chicken breast can also be cooked through more easily.

Braised Dried Abalone with Goose's Web

*Chinese version refer to p.32/makes 1 serving

INGREDIENTS:

1 abalone
1 goose web
2 leafy green stems

BRAISING STOCK FOR GOOSE WEB: (FOR 100 GOOSE WEBS)

8 slices ginger
8 sprigs spring onion
150 g garlic cloves
75 g shallot
1.2 kg fatty pork
900 g bones of Jinhua ham

SEASONING FOR GOOSE WEB:

dark soy sauce
water (enough to cover the goose webs)
225 g oyster sauce

BRAISING STOCK FOR ABALONE: (FOR 3 KG OF DRIED ABALONES)

2 chicken (about 3.6 kg)
1.2 kg lean pork
1.2 kg pork belly ribs
600 g chicken feet
pork skin
10 dried scallops
spring onion

SEASONING FOR ABALONE:

20 g rock sugar
dark soy sauce

METHOD:

BRAISING DRIED ABALONE:

1. Soak dried abalones in water until soft, for about 1 week. The soaking time depends on their sizes.
2. Steam abalones over boiling water until soft, for about 8 hours. The steaming time depends on their sizes. Set aside.
3. Rinse chicken and chop into pieces. Rinse lean pork and pork belly ribs. Cut into chunks. Rinse chicken feet and pork skin. Soak dried scallops in water until soft.
4. Heat oil in a wok. Stir-fry spring onion until fragrant. Put in chicken, lean pork, pork belly ribs, chicken feet and pork skin. Fry until fragrant. Set aside.
5. Line the bottom of a claypot or ceramic pot with a bamboo mat. Arrange ingredients from step 4 over the bamboo mat. Add dried scallops, abalones and seasoning. Bring to the boil and turn to low heat. Simmer for 3 hours until abalones are tender. Save abalones on a serving plate. Remove all solid ingredients and strain the sauce. Set aside.

BRAISING GOOSE WEB:

1. Rinse the goose webs. Add some dark soy sauce and mix well to colour them evenly.
2. Heat oil in a wok until smoky. Deep-fry the goose webs for 30 seconds until golden. Set aside.
3. Stir-fry ginger, spring onion, garlic cloves and shallot in a wok. Add fatty pork and bone of Jinhua ham. Add seasoning. Bring to the boil and turn to low heat. Simmer for 75 minutes. Turn off the heat and leave the lid on. Leave them for 90 minutes until tender.
4. Arrange on the same plate as the abalone. Garnish with blanched leafy green stems. Serve hot.

MASTER FU'S TIPS:

- For this recipe, Master Fu chooses goose webs from Poland, because of their hefty sizes and velvety texture.
- When you colour the goose webs with dark soy sauce, pick a brand that isn't too dark in colour. Otherwise, the dish won't look as good.
- When you braise goose webs, you need to add fatty pork to the braising stock, so as to elevate the aromas.

Stuffed Bamboo Piths with Bird's Nest

*Chinese version refer to p.36/makes 1 serving

INGREDIENTS:

1 bamboo pith (about 7 to 8 cm long, choose a thicker one)
38 g bird's nest
2 leafy green stems
premium stock (enough to cover the bird's nest)
shredded cooked Jinhua ham (for decoration)

METHOD:

1. Soak bamboo pith in water for 30 minutes until soft. Cut off both ends. Blanch in boiling water and drain well. Simmer in premium stock to infuse it with flavours. Drain and set aside.
2. Rehydrate the bird's nest. Drain. Transfer into a steaming bowl and add premium stock to cover. Steam for 12 minutes. Drain. Stir in thickening glaze and cook till it thickens. Set aside.
3. Stuff the bamboo pith with the bird's nest. Arrange on a serving dish. Garnish with leafy green stems. Drizzle with the thickened stock. Garnish with shredded cooked Jinhua ham. Serve.

MASTER FU'S TIPS:

The bamboo pith and bird's nest will soak up the flavour of the premium stock for rounded, mellow umami. It's worth spending the time and energy to make your own premium stock with ingredients like old chicken and Jinhua ham. You can taste the difference.

Wok-fried Shark's Fin with Fresh Crabmeat, Egg and Bean Sprouts

*Chinese version refer to p.38

INGREDIENTS:

120 g shark's fin (rehydrated)
38 g crabmeat
58 g mung bean sprouts (both ends cut off)
20 g shredded char siu pork
shredded Jinhua ham
2 large eggs
yellow chives
premium stock (for simmering the shark's fin)

SEASONING:

salt, caltrop starch slurry, Shaoxing wine, sesame oil

METHOD:

1. Rinse the shark's fin. Steam until soft. Blanch it in boiling water with ginger and spring onion to remove the fishy smell. Drain and transfer into a double-boiling pot. Add premium stock to cover. Double-boil until cooked through. Drain and set aside.
2. Blanch mung bean sprouts in premium stock until medium-well done. Drain and set aside. Whisk the eggs and season with a pinch of salt.
3. Heat oil in a wok. Stir-fry the egg quickly into tiny bits that resemble osmanthus flowers. Add shark's fin and crabmeat. Toss until fragrant. Add char siu pork, mung bean sprouts and yellow chives. Toss again to mix well. Drizzle with Shaoxing wine, sesame oil and caltrop starch slurry. Toss again. Garnish with shredded Jinhua ham at last. Serve.

MASTER FU'S TIPS:

You have to be patient when preparing shark's fin. It is the hero of the whole dish, so that it's worth taking the time and effort to simmer it in premium stock slowly, for the flavours to build up and for it to turn soft and velvety. That's how you elevate this Cantonese classic to the next level.

Crispy Fried Chicken

*Chinese version refer to p.40

INGREDIENTS:

1 dressed chicken (about 1.8 kg)

15 g salt

BASTING SAUCE:

200 g white vinegar

50 g maltose

5 g red vinegar

5 g double-distilled rice wine

METHOD:

1. To make the basting sauce, put white vinegar, maltose and red vinegar into a bowl. Put the bowl into a boiling water bath until the maltose dissolves. Leave it to cool. Stir in double-distilled rice wine. Mix well.
2. Rinse the chicken and remove the innards. Rinse well and wipe dry.
3. Rub salt on the skin of the chicken. Leave it for 4 hours.
4. Boil a pot of water. Hold the chicken by its neck and dip the chicken in the boiling water until the skin tightens up. Hang it to air dry. Brush on the basting sauce evenly all over the skin. Hang it up again to air dry for 6 hours until dry to touch.
5. Heat oil in a wok up to 120°C. Put in the chicken and deep-fry for 10 minutes. Turn the heat up to 160°C and deep-fry until the skin is crispy and golden. Remove from the oil and slice it. Serve.

MASTER FU'S TIPS:

- Master Fu only uses freshly slaughtered three-yellow chicken for this recipe.
- To make the chicken skin crispy while keeping the flesh juicy, the key is the control the heat well. Soak the chicken in oil at 120°C until the meat is just cooked through. Then turn the heat up to 160°C to crisp up the skin.

Braised "Buddha Jumps over the Wall"

*Chinese version refer to p.42

INGREDIENTS:

1 Japanese Yoshihama abalone (about 17 to 18 g before rehydration)
40 g rehydrated fish maw
60 g sea cucumber
20 g shark's fin
80 g abalone sauce

BRAISING STOCK FOR ABALONE: (FOR 3 KG OF DRIED ABALONE)

2 fresh chicken (about 3.6 kg)
1.2 kg lean pork
1.2 kg pork belly ribs
600 g chicken feet
pork skin
10 dried scallops
spring onion

SEASONING FOR ABALONES:

20 g rock sugar
dark soy sauce

METHOD:

BRAISING ABALONE:

1. Soak dried abalone in water until soft, for about 1 week. The soaking time depends on their sizes.
2. Steam the abalones over boiling water till soft, for about 8 hours. The steaming time depends on their sizes. Set aside.
3. Rinse the chicken and chop into pieces. Rinse the lean pork and pork belly ribs. Cut into chunks. Rinse chicken feet and pork skin. Soak dried scallops in water until soft.
4. Heat oil in a wok. Stir-fry spring onion until fragrant. Put in chicken, lean pork, pork belly ribs, chicken feet and pork skin. Fry until fragrant. Drain and set aside.
5. Line the bottom of a clay pot with a bamboo mat. Arrange the fried ingredients from step 4 over the bamboo mat. Add dried scallops, abalones and seasoning. Bring to the boil and turn to low heat. Simmer for 3 hours until abalones are tender. Remove all solid ingredients and strain the sauce for later use.

REHYDRATING FISH MAW:

Boil a pot of hot water. Soak dried fish maw in hot water until it turns cold. Leave it to soak until fully rehydrated (usually for a few days). Boil a pot of second stock and add ginger, spring onion and deep-fried garlic cloves. Put in the fish maw and simmer to remove any fish smell. Drain and set aside.

REHYDRATING SEA CUCUMBER:

1. Soak dried sea cucumber in water for 1 to 2 days. Boil a pot of water and turn off the heat. Put in the sea cucumber and cover the lid. Leave it in the water overnight until the water turns cold.
2. Cut open the belly of the sea cucumber. Remove all sand and innards. Boil a pot of water and turn off the heat. Put in the sea cucumber and cover the lid. Leave it in the water till the water turns cold. Repeat this soaking step until the sea cucumber is soft. Soak it in cold water.
3. Boil a pot of second stock. Add ginger, spring onion and garlic cloves. Leave the stock to cold. Soak the sea cucumber in the cold stock to pick up flavours. Then boil a pot of premium stock. Turn to low heat. Put in the sea cucumber and simmer briefly. Set aside.

PREPARING SHARK'S FIN:

Rehydrate the shark's fin. Rinse well and steam until soft. Cook shark's fin in water with ginger and spring onion to remove fish smell. Then transfer into a double-boiling pot of premium stock. Double-boil in a boiling water bath over low heat until flavourful. Set aside.

ASSEMBLY:

Make sure all ingredients are braised till soft. Bring the abalone sauce to the boil. Put in the shark's fin, abalone, sea cucumber and fish maw to heat them through. Stir in thickening glaze. Save on a serving plate and serve.

MASTER FU'S TIPS:

Every component of this gourmet dish need to be prepared meticulously - including the soaking and braising time. That's how you can make the perfect "Buddha Jumps over the Wall"

Celebrity Shark's Fin in Premium Stock Glaze

*Chinese version refer to p.48/makes 1 serving

INGREDIENTS:

180 g shark's fin
shredded Jinhua ham
sliced ginger
spring onion (cut into short lengths)

BRAISING STOCK FOR SHARK'S FIN:

Shaoxing wine
80 g concentrated chicken stock
caltrop starch slurry

ON THE SIDE:

mung bean sprouts (both ends cut off)
yellow chives
1 bowl premium stock

METHOD:

1. Rehydrate the dried shark's fin. Rinse well. Steam until soft. Blanch briefly in water with spring onion and ginger to remove the fishy smell.
2. Heat oil in a wok. Sprinkle with Shaoxing wine. Add concentrated chicken stock. Bring to the boil. Add shark's fin and simmer until it picks up the flavour. Arrange shark's fin on a serving dish. Thicken the stock with caltrop starch slurry. Cook while stirring until thick. Drizzle over the shark's fin. Garnish with shredded Jinhua ham.
3. Serve with blanched mung bean sprouts, yellow chives and a bowl of premium stock on the side.

MASTER FU'S TIPS:

For the best result, make sure you control the heat and time when you blanch, simmer and double-boil the shark's fin. Do not overcook it. Otherwise, it will be too soggy without the slippery but still toothsome texture.

Double-boiled Shark's Fin with Yunnan Ham and Leafy Green Stems

*Chinese version refer to p.50/makes 1 serving

INGREDIENTS:

80 g shark's fin
2 leafy green stems
1 slice Yunnan ham
240 g premium stock
sliced ginger
spring onion (cut into short lengths)

PREMIUM STOCK:

1 old chicken
480 g lean pork
160 g pork leg bone
140 g beef
140 g Jinhua ham
4.8 litres water

METHOD:

1. Rinse and clean the ingredients for the premium stock. Put them into a large stock pot. Bring to the boil and simmer for 2 to 3 hours until the liquid reduces to 2.4 litres. For this recipe, you need 240 g of this stock.
2. Blanch the leafy green stems in boiling water. Soak them in premium stock until flavourful. Drain.
3. Rinse the shark's fin. Sandwich it between two bamboo mats. Steam over high heat for 6 hours until soft.
4. Cook the shark's fin in water with ginger and spring onion to remove fishy smell.
5. Transfer shark's fin into a double-boiling pot. Add Yunnan ham and leafy green stems. Pour in premium stock. Double-boil for 2 hours in a boiling water bath. Serve.

MASTER FU'S TIPS:

Master Fu suggests sandwiching the shark's fin in two bamboo mats before steaming it. That would make the shark's fin soft, while keeping its shape for a striking presentation.

Pan-fried Shark's Fin in Sour Broth

*Chinese version refer to p.52

INGREDIENTS:

1 shark's fin
premium stock
sliced ginger
spring onion (cut into short lengths)

SOUR BROTH:

1.8 kg celery, onion, green and red bell pepper, carrot, tomato, salted mustard green
1.6 kg assorted fish
1.6 kg lean pork
Sichuan peppercorns
rice vinegar
7.2 litres water

METHOD:

1. Rinse all sour broth ingredients. Peel them if necessary. Cut into chunks. Put all ingredients into a stock pot. Bring to the boil and simmer until the liquid reduces to 3.6 litres.
2. Rehydrate the dried shark's fin. Rinse well. Sandwich the shark's fin between two bamboo mats. Steam until soft. Blanch it in water with ginger and spring onion to remove the fishy smell.
3. Transfer the shark's fin into a boiling premium stock. Simmer until it picks up the flavour. Set aside.
4. Heat oil in a wok. Fry the shark's fin until both sides golden. Arrange on a serving plate.
5. Put a ladle of the sour broth from step 1 into a small pot. Stir in caltrop starch slurry. Cook until it thickens. Drizzle over the shark's fin. Serve.

MASTER FU'S TIPS:

- For this recipe, Master Fu uses Ya Jian shark's fin, which is a medium-quality item from the shark's dorsal fin. It has fine and dense cartilage that boasts a slippery and velvety texture. It picks up the sour broth flavour nicely.
- You need patience when pan-frying shark's fin. If the shark's fin is too wet and your wok is not hot enough, it tends to stick. Make sure the wok is heated hot enough before putting in the shark's fin.

Red Braised Giant Fish Maw

*Chinese version refer to p.54

INGREDIENTS:

1 Indian cod fish maw (about 150 g)
320 g premium stock (or abalone sauce)
sliced ginger
spring onion (cut into short lengths)
garlic cloves
1.14 kg premium stock
6 iceberg lettuce stems

METHOD:

1. Rehydrate the dried fish maw. Boil a pot of water and turn off the heat. Put in the fish maw. Cover the lid and leave it till the water turns cold. Repeat the boiling and soaking steps until the fish maw is soft.
2. Blanch the fish maw in premium stock with ginger, spring onion and garlic cloves to remove the fishy smell. Transfer into a pot of premium stock and cook until flavourful and soft. Transfer onto a serving dish.
3. Bring the premium stock or abalone sauce to the boil. Stir in caltrop starch slurry and cook till it thickens. Drizzle over the fish maw. Arrange blanched lettuce leaves on the side.

MASTER FU'S TIPS:

Cod fish maw is usually produced in Indonesia or Pakistan. It has firm texture and is thicker than its counterparts. It is considered the premium choice in fish maw. To make good use of such luxurious ingredients, one must control the soaking and braising time precisely.

Braised Dried Sea Cucumber in Abalone Sauce

*Chinese version refer to p.56/makes 1 serving

INGREDIENTS:

1 dried spiny sea cucumber
abalone sauce
premium stock
2 leafy green stems

BRAISING STOCK:

sliced ginger
garlic cloves
spring onion (cut into short lengths)
premium stock

METHOD:

1. Soak the dried sea cucumber in water for 1 to 2 days. Boil a pot of water and turn off the heat. Put in the sea cucumber and cover the lid. Leave it in the water overnight.
2. Cut open the ventral side of the sea cucumber. Remove the sand and innards. Boil a pot of water and turn off the heat. Put in the sea cucumber and leave it till the water turns cold. Repeat this heating and soaking steps until soft. Soak it in cold water for later use.
3. Put ginger, garlic cloves and spring onion into premium stock at room temperature. Soak the sea cucumber in it until flavourful. Then bring the stock to the boil and turn to low heat. Simmer the sea cucumber.
4. Bring the abalone sauce to the boil. Put in the sea cucumber and cook briefly. Stir in caltrop starch slurry and cook till it thickens. Save the sea cucumber and the sauce on a serving dish. Arrange blanched leafy green stems on the side. Serve.

MASTER FU'S TIPS:

Master Fu chooses spiny sea cucumber from Hokkaido for this recipe. Such sea cucumbers live in pristine, cold sea water all year round. They taste velvety, but toothsome. They are the gift of nature, with exceptional nutritional value.

Bird's Nest in Chicken Juices

*Chinese version refer to p.58/makes 1 serving

INGREDIENTS:

80 g bird's nest
1 dressed chicken (about 1.8 kg)
premium stock
edible gold leaves

METHOD:

1. Soak the bird's nest in water overnight. Drain well. Simmer it in premium stock until flavourful and soft. Drain again.
2. Rinse the chicken and remove all innards. Rinse again. Put it in a steaming bowl and steam for 3 hours. Collect the juices that come out of the chicken in the bowl. You need 80 g for this recipe.
3. Bring the chicken juices to the boil and stir in caltrop starch slurry. Cook until it thickens into a thin glaze. Pour it into a serving dish. Arrange the cooked bird's nest on top. Garnish with edible gold leaves.

MASTER FU'S TIPS:

The chicken juices are made with steaming a whole chicken, without adding any water. They are pure essence. This recipe lets you savour the velvety texture of bird's nest and the rounded flavour of the chicken juices.

Bird's Nest and Partridge Porridge

*Chinese version refer to p.60/makes 1 serving

INGREDIENTS:

80 g bird's nest
25 g chopped partridge meat
60 g grated yam
shredded Jinhua ham
80 g premium stock

METHOD:

1. Soak the bird's nest in water overnight. Simmer it in premium stock until flavourful and soft. Drain and set aside.
2. Rinse the partridge. Remove the innards. Rinse well and fillet the breast.
3. Remove the tough membrane on the filleted breast. Finely chop it.
4. Bring premium stock to the boil. Put in the chopped partridge and grated yam. Cook until creamy. Stir in caltrop starch slurry and cook until thick.
5. Transfer the porridge into a serving dish. Arrange bird's nest on top. Sprinkle with shredded Jinhua ham. Serve.

MASTER FU'S TIPS:

For the best result, make sure you remove the tough membrane on the partridge breast, and finely chop the meat. That would make the porridge silky and creamy. In well-off families, this is a health tonic for children.

Double-boiled Bird's Nest with Crystal Sugar

*Chinese version refer to p.62/makes 1 serving

INGREDIENTS:

80 g bird's nest
rock sugar
80 g water
evaporated milk
coconut milk
edible gold leaves

METHOD:

1. Soak dried bird's nest in water overnight. Drain well.
2. Put bird's nest into a double-boiling pot. Add rock sugar and water. Double-boil in a boiling water bath until rock sugar dissolves. Drain and transfer the bird's nest into a serving bowl.
3. Pour in evaporated milk and coconut milk. Garnish with edible gold leaves. Serve.

MASTER FU'S TIPS:

Double-boiling bird's nest with rock sugar lets the bird's nest soak up the sweetness of rock sugar. Then drain all liquid so that the sweet soup will not be too watery. This Cantonese sweet soup lets you taste the silky texture of bird's nest, while tonifying your lungs and strengthening your body.

Fried Chicken Skin Stuffed with Shrimp Mousse

*Chinese version refer to p.66

INGREDIENTS:

1 chicken

500 g minced filling

MINCED SHRIMP FILLING:

1.2 kg medium marine shrimps

20 g fatty pork (diced)

1 egg white

4 g salt

4 g sugar

12 g caltrop starch

BASTING SAUCE:

200 g white vinegar

50 g maltose

5 g red vinegar

5 g double-distilled rice wine

METHOD:

MINCED SHRIMP FILLING:

1. Shell and devein the shrimps. Rub salt and caltrop starch on them. Rinse well. Wipe dry with a clean dry towel.
2. Put the shrimps on a chopping board. Crush the shrimps with the flat side of a cleaver. Coarsely chop them with the back of knife.
3. Put the minced shrimps into a mixing bowl. Add egg white and stir in one direction with a bare hand until sticky. Add salt, sugar, caltrop starch and diced fatty pork. Mix well. Lift the minced shrimp mixture off the bowl and slab it back into the bowl repeatedly until sticky and resilient. This is the filling.

BASTING SAUCE:

Mix white vinegar, maltose and red vinegar in a bowl. Put it in a boiling water bath and cook until the maltose dissolves. Leave it to cool. Add double-distilled rice wine and mix well.

ASSEMBLY:

1. Rinse the chicken and carefully remove all bones and meat (leave the wings on) without damaging the skin.
2. Secure the chicken skin on a bamboo mat with metal skewers. Pour hot water over it a few times until the skin no longer shrinks and holds its shape.
3. Brush the basting sauce over the chicken skin. Hang it up to air-dry.
4. Coat the insides of the skin with caltrop starch. Spread the minced shrimp filling evenly over it. Deep-fry in hot oil until golden and cooked through. Slice and serve.

MASTER FU'S TIPS:

Make sure you coat the insides of the chicken skin with caltrop starch evenly before stuffing it with minced shrimp filling. The caltrop starch acts like adhesive to stick the minced shrimp filling on the chicken skin, keeping it securely attached after deep-frying.

Master Fu's Favourite Sweet and Sour Pork

*Chinese version refer to p.70

INGREDIENTS:

113 g pork collar
160 g onion, red and green bell pepper
6 pieces fresh pineapple
sweet and sour sauce

AROMATICS:

grated ginger
2 tsp grated garlic
spring onion (cut into short lengths)

METHOD:

1. Slice the pork collar thickly, each slice about 3 cm by 1 cm and 3 mm thick. Rinse well and wipe dry.
2. Deseed the bell peppers. Cut into wedges and blanch in boiling water. Cut onion and pineapple into wedges.
3. Dip the pork collar into a thin deep-frying batter. Then coat it in caltrop starch. Deep-fry in hot oil until crispy. Drain and let it sit for a couple minutes. Put them in hot oil to fry for the second time. Drain.
4. Heat oil in a wok. Stir-fry aromatics and onion until fragrant. Add the sweet and sour sauce and cook till it boils. Stir in caltrop starch slurry and toss till it thickens. Put the deep-fried pork collar in. Toss well to coat evenly. Add bell peppers and pineapple. Toss again. Serve.

MASTER FU'S TIPS:

- There is only 450 g of pork collar meat in each pig – it's considered a rare cut. Master Fu slices it to shorten the deep-frying time. The pork will end up crispy and bouncy in texture, quite different from the typical sweet and sour pork.
- To make the sweet and sour sauce, put sugar, raw cane sugar slab, white vinegar, ketchup, OK sauce, Worcestershire sauce, celery, coriander, red and green bell pepper, and red chilli into a pot. Cook until the veggies are soggy. Strain the liquid.

Baked Crab Shell Stuffed with Fresh Crabmeat, Onion with Cream Sauce

*Chinese version refer to p.72/makes 1 serving

INGREDIENTS:

32 g crabmeat (freshly picked from a steamed swimmer crab)
48 g shredded onion
butter
40 g whipping cream
40 g premium stock
mashed potato
1 egg yolk
grated cheddar
crab shell

METHOD:

1. Heat oil in a wok. Put in 10 g of crabmeat and onion. Stir until fragrant. Add butter, whipping cream and premium stock. Mix well.
2. Stuff the crab shell with the remaining 22 g of crabmeat. Top with the onion mixture from step 1.
3. Smear a layer of mashed potato on top of the onion filling. Brush with whisked egg yolk. Sprinkle with grated cheddar. Bake in a preheated oven at 220°C for 12 minutes until golden. Serve.

MASTER FU'S TIPS:

- Instead of mixing all crabmeat with the onion, Master Fu stuffs the crab shell with 22 g of crabmeat first, before topping it with the onion mixture. That would ensure your guests get to savour intense crab flavour with every bite.
- When you make the crabmeat and onion filling, try to hold back on adding water or over-seasoning. You don't want it to be watery, but you don't want it to overpower the delicate crabmeat flavour either.

Pan-fried Phoenix Shrimp

*Chinese version refer to p.74/makes 1 serving

INGREDIENTS:

1 large shrimp (about 50 to 60 g)
40 g minced shrimp filling
caltrop starch
salt
spring onion (finely chopped)

MINCED SHRIMP FILLING:

1.2 kg medium marine shrimps
20 g diced fatty pork
1 egg white
4 g salt
4 g sugar
12 g caltrop starch

THICKENING GLAZE:

10 g butter
60 g premium stock
salt
caltrop starch slurry

METHOD:

MINCED SHRIMP FILLING:

1. Shell and devein the shrimps. Rub salt and caltrop starch on them. Rinse well. Wipe dry with a clean dry towel.
2. Put the shrimps on a chopping board. Crush the shrimps with the flat side of a cleaver. Coarsely chop them with the back of knife.
3. Put the minced shrimps into a mixing bowl. Add egg white and stir in one direction with a bare hand until sticky. Add salt, sugar, caltrop starch and diced fatty pork. Mix well. Lift the minced shrimp mixture off the bowl and slab it back into the bowl repeatedly until sticky and resilient. This is the filling.

ASSEMBLY:

1. Remove the head of the large shrimp. Shell it but keep the tail intact. Rub salt and caltrop starch on it. Rinse well. Wipe dry with a clean dry towel.
2. Make a cut along the back of the shrimp. Devein. Pound it gently to flatten it slightly. Then make a few light cuts on the belly side.
3. Coat the large shrimp lightly in caltrop starch. Spread the minced shrimp filling generously and evenly on the back of the large shrimp.
4. Heat oil in a pan. Fry the stuffed shrimp with the filling side down over low heat. Flip it to fry the other side until medium-well done. Save on a serving plate.
5. Melt butter in a pot. Add premium stock and bring to the boil. Season with salt. Stir in caltrop starch slurry and cook until it thickens. Drizzle over the fried stuffed shrimp. Sprinkle with finely chopped spring onion. Serve.

MASTER FU'S TIPS:

Making light cuts on the belly side of the shrimp helps make it curl less when cooked. That would ensure the best adhesion between the shrimp and the minced shrimp filling.

Home Style Steamed Giant Grouper

*Chinese version refer to p.78

INGREDIENTS:

6 pieces of giant grouper fillet (about 38 g each)
30 g salted kohlrabi (finely shredded)
3 slices red chilli
spring onion (finely chopped)
shredded ginger
shredded shiitake mushroom
shredded dried tangerine peel

MARINADE:

cooking oil
salt
caltrop starch

AROMATICS:

grated garlic
chopped shallot
finely diced ginger

METHOD:

1. Rinse the grouper fillet. Wipe dry. Add marinade and mix well.
2. Heat oil in a wok and stir-fry aromatics until fragrant.
3. Arrange grouper fillet on a steaming plate. Sprinkle with kohlrabi, ginger, shiitake mushrooms, dried tangerine peel, red chilli and the fried aromatics from step 2.
4. Boil water in a wok or steamer. Steam the grouper fillet over high heat for 4 minutes. Turn off the heat and leave it in with the lid cover for 1 minute. Sprinkle with spring onion and pour sizzling hot oil over it. Drizzle with soy sauce to taste. Serve.

MASTER FU'S TIPS:

- Salted kohlrabi from Shunde is meticulously preserved, imparting crispy texture and briny umami. It is best paired with meat or seafood. From Chinese medical point of view, it tonifies the Spleen, whets the appetite, moistens the lungs and alleviates coughing.
- This recipe pair grouper fillet with salted kohlrabi, dried tangerine peel, shiitake mushrooms and ginger. The condiments are aromatic and salty in taste, removing the fishy smell of the fish. It also imparts rustic, farmhouse charm.

Stuffed Eight Treasure Duck (Whole Duck)

*Chinese version refer to p.80

INGREDIENTS:

1 whole duck
4 slices ginger
2 sprigs spring onion
3 cloves star-anise
premium stock

STUFFING:

6 salted egg yolks (diced)
30 g lotus seeds
30 g lily bulbs
30 g pearl barley
80 g shelled chestnuts
30 g gingkoes
10 g diced shiitake mushrooms
10 g diced Jinhua ham
30 g glutinous rice

COLOURING FOR DUCK SKIN:

dark soy sauce

METHOD:

1. Rinse the dressed duck and remove all innards. Debone it carefully without damaging the skin.
2. Steam glutinous rice until done. Heat oil in a wok and stir-fry remaining stuffing ingredients. Add cooked glutinous rice and mix well. Leave the stuffing to cool. Stuff the deboned duck with the stuffing. Seal the seam using a trussing needle and butcher's twine. Brush dark soy sauce evenly over the skin.
3. Heat oil in a wok. Deep-fry the stuffed duck on all sides to keep its shape. Drain.
4. Put the duck into a large dish. Add ginger, spring onion, star-anise and premium stock. Steam over high heat for 90 minutes. Thicken the sauce with caltrop starch slurry. Serve with blanched leafy green stems on the side.

MASTER FU'S TIPS:

- This traditional painstaking recipe is complicated with many steps. The hardest part is probably deboning the duck without damaging the skin. It takes a few hours for the preparation and cooking steps.
- Master Fu uses smaller grain-fed ducks for this recipe - they have less fat and their meat is succulent and velvety. They offer layers of textures and flavours.
- Lily bulbs and chestnuts are added for their medicinal value - they benefit the Spleen, promote Stomach functions, calm the nerves, and nourish the body.
- The glutinous rice and eight-treasure stuffing pick up the essence and juices of the duck nicely. The flavours are rounded and rich; the texture is soft and chewy.

Live Shrimp with Sweet and Sour Sauce

*Chinese version refer to p.84/makes 1 serving

INGREDIENTS:

1 large shrimp (about 85 to 100 g)
1 piece fresh pineapple
whisked egg
caltrop starch
sweet and sour sauce (see p.162 for method)

METHOD:

1. Remove the sand sac in the shrimp's head. Then remove the shell while keeping the head and the tail intact. Make a cut along the back of the shrimp. Rinse well.
2. Dip the shrimp in whisked egg and coat it in caltrop starch. Deep-fry in hot oil until cooked through and golden. Save on a serving plate.
3. Heat the sweet and sour sauce. Drizzle over the shrimp. Garnish with pineapple. Serve.

MASTER FU'S TIPS:

Shelling the shrimp partly and butterflying it helps it pick up the sweet and sour sauce readily. Master Fu keeps the head and tail intact for better presentation.

Crab-filled Medallions in Daliang Style

*Chinese version refer to p.86

INGREDIENTS:

1 slice fatty pork (about 8 cm X 8 cm X 0.5 mm)
crab roe
coriander leaves

FILLING:

320 g diced pork tenderloin
320 g diced shrimps
400 g diced bamboo shoot
240 g crabmeat
120 g diced yellow chives

EGG WHITE WASH:

2 egg whites (whisked)
corn starch
water

METHOD:

1. Slice the fatty pork thinly. Cut into round discs using a cookie cutter.
2. Mix the filling ingredients. Lay flat a disc of fatty pork on the counter. Put 5 g of filling over it. Add a coriander leaf and some crab roe. Top with another disc of fatty pork. Press well to secure the seam. Repeat this step until all filling is used up.
3. Mix the egg white wash ingredients. Dip the stuffed medallions into the egg white wash. Deep-fry in oil until crispy and fluffy. Serve.

MASTER FU'S TIPS:

- Make sure you control the oil temperature and deep-frying time well. Do not deep-fry over high heat. The medallions should look golden, tasting crispy without being greasy.
- If you find it hard to slice the fatty pork with a knife, put the pork in the freezer until firm, then use a deli meat slicer instead.

Braised Pomelo Skin with Shrimp Roe

*Chinese version refer to p.88

INGREDIENTS:

20 pieces pomelo skin
shrimp roe

BRAISING STOCK:

800 g dace
400 g pork ribs
200 g lean pork
60 g dried shrimps
bones of Jinhua ham
skin of Jinhua ham
ginger
spring onion

SEASONING:

oyster sauce
dark soy sauce

METHOD:

1. Peel off the green zest of the pomelo skin. Use only the white part for this recipe. Blanch the pomelo skin in boiling water. Rinse in running tap water. Squeeze dry. Soak it in water for 1 to 2 days to remove its bitter taste. Wipe dry.
2. Heat oil in a wok. Deep-fry the pomelo skin until stiff. Drain.
3. Rinse the dace and remove all innards. Rinse pork ribs and lean pork. Wipe dry.
4. Heat oil in a wok. Deep-fry dace, pork ribs and lean pork until golden. Drain and set aside.
5. In the same wok, stir-fry ginger, spring onion and dried shrimps until fragrant. Put in the pomelo skin and all braising stock ingredients. Add water and bring to the boil. Turn to low heat and simmer for 90 minutes until the pomelo skin is soft and flavourful.
6. Arrange on a serving plate. Sprinkle with shrimp roe. Serve.

MASTER FU'S TIPS:

This is a complicated recipe with many steps. You should especially pay attention to the soak step of the pomelo skin. You need to rinse it under running tap water and change the water repeatedly to remove its bitter taste. Otherwise, the dish will turn out bitter after simmered.

Crispy Baby Pigeon

*Chinese version refer to p.90

INGREDIENTS:

1 baby pigeon
maltose
water

ACCOMPANIMENTS:

five-spice powder
Worcestershire sauce

METHOD:

1. Remove the innards of the baby pigeon. Rinse well. Wipe dry.
2. Put maltose and water into a bowl. Put the bowl into a hot water bath until maltose dissolves. Set aside.
3. Marinate the pigeon with a special blend of soy sauce. Brush diluted maltose over the skin. Hang it up to air-dry. Set aside.
4. Heat oil in a wok. Put in the pigeon with the back side down. Then flip to fry the breast. Deep-fry until the pigeon is evenly browned on all sides. Save on a serving plate. Sprinkle with five-spice powder and/or Worcestershire sauce to elevate its flavour. Serve.

MASTER FU'S TIPS:

- The back and wings of the pigeon are harder to cook through. That's why you should fry the back side first before flipping it to fry the breast. That would keep the bird crispy on the outside, juicy on the inside. You're less likely to overcook the breast that way.
- Another trick is to put the pigeon in hot oil upside-down. Its wings are spread when put upside-down, so that it takes less time to cook in oil, keeping the meat moist.

Deep-fried Shrimp on Toast

*Chinese version refer to p.92

INGREDIENTS:

6 to 8 medium shrimps (about 38 g each)
4 slices sandwich bread
6 to 8 coriander leaves
shredded Jinhua ham

EGG WHITE WASH

2 egg whites
corn starch
water

METHOD:

1. Shell the shrimps. Rinse well. Make a cut along the back. Devein.
2. Cut the sandwich bread into rectangular shapes. Brush some egg white wash on it. Spread a shrimp flat and put over the bread. Press to secure. Put a coriander leaf and some shredded Jinhua ham on top.
3. Heat oil in a wok. Deep-fry the shrimp on toast until golden and cooked through. Drain. Save on a serving plate and serve.

MASTER FU'S TIPS:

- Egg white wash is a good adhesive to stick the shrimp to the bread. It helps sticking the shrimp firmly on the bread and reduces the chance of them falling apart.
- Control the oil temperature well. Do not fry over high heat. Otherwise, the toast will be browned too much and doesn't look appetising.

Stuffed Pork Intestine with Cuttlefish and Black Truffle

*Chinese version refer to p.94

INGREDIENTS:

1 pork intestine
ginger
spring onion
star-anise
bay leaves
salt
sweet and sour sauce (see p.162 for method)

CUTTLEFISH FILLING:

250 g cuttlefish
1 black truffle (finely chopped)
20 g diced fatty pork
egg white
salt
caltrop starch

BASTING SAUCE:

200 g white vinegar
50 g maltose
5 g red vinegar
5 g double-distilled rice wine

METHOD:

1. Rub the pork intestine clean thoroughly. Rinse well.
2. Boil a pot of water. Add ginger, spring onion, star-anise, bay leaf and salt. Turn the pork intestine inside out and boil it in water for 1 hour. Skim off any visible fat and grease on top. Turn off the heat. Cover the lid and leave the pork intestine in the water for 15 minutes. Set aside.
3. To make the cuttlefish filling, rinse the cuttlefish. Drain and cut into chunks. Add the remaining cuttlefish filling ingredients. Transfer into a food processor. Grind the mixture into a thick paste. Transfer into a mixing bowl. Lift the mixture off the bowl and slap it back forcefully into the bowl repeatedly until resilient and sticky.
4. Stuff the pork intestine with the cuttlefish filling. Brush the basting sauce on the pork intestine. Hang it up to air-dry.
5. Deep-fry the stuffed pork intestine in hot oil until golden and crispy. Slice at an angle. Serve with sweet and sour sauce on the side.

MASTER FU'S TIPS:

The key to success is to control the oil temperature and deep-frying time precisely. That's how you can make the stuffed pork intestine crispy on the outside, bouncy and juicy on the inside.

Honey Glazed Semi-dried Oysters

*Chinese version refer to p.98

INGREDIENTS:

6 to 8 semi-dried oysters

SAUCE:

20 g honey
10 g sugar
light soy sauce
dark soy sauce
20 g premium stock

METHOD:

1. Rinse the semi-dried oysters. Wipe dry.
2. Boil water in a wok or steamer. Steam the semi-dried oysters until cooked through. Heat oil in a pan. Fry the semi-dried oysters until lightly browned. Set aside.
3. Bring the sauce ingredients to the boil. Put the pan-fried oyster back in. Toss to coat well. Serve.

MASTER FU'S TIPS:

- The Chinese believe oysters boost the vital essence in human, while tonifying the Kidneys. Master Fu uses semi-dried oysters farmed in the South China Sea. They boast strong seafood umami, with distinctive aromas.
- The sweet and rich honey glaze elevates the flavours of semi-dried oysters in major ways.

Wok-fried Fresh Crabmeat with Egg, Conpoy and Bean Sprouts

*Chinese version refer to p.100

INGREDIENTS:

3 eggs
40 g rehydrated conpoy
60 g mung bean sprouts (with both ends cut off)
80 g crabmeat
20 g char siu pork
spring onion (finely chopped)
yellow chives (added at last)
shredded cooked Jinhua ham

SEASONING:

salt
Shaoxing wine
sesame oil

METHOD:

1. Squeeze dry the rehydrated conpoy. Deep-fried half of it in warm oil until golden. Drain and set aside.
2. Blanch the mung bean sprouts in premium stock until medium-well done. Drain and set aside. Whisk the eggs and season them with salt.
3. Heat a wok until red hot. Pour in cooking oil. Then add whisked eggs, crabmeat, rehydrated conpoy, char siu pork and chopped spring onion. Toss quickly until the eggs are set into small bits like osmanthus flowers. Add mung bean sprouts and yellow chives. Toss again. Season with salt, Shaoxing wine and sesame oil. Save on a serving plate. Garnish with deep-fried conpoy and shredded cooked Jinhua ham. Serve.

MASTER FU'S TIPS:

The fried egg bits should resemble osmanthus flowers, but without being dry and rubbery. The bean sprouts should be cooked, but still crisp. This is a recipe that builds layers of textures.

Lobster with Shrimp Ball Stuffed with Crab Roe

*Chinese version refer to p.102

INGREDIENTS:

1 lobster (about 750 g)
2 female mud crabs
chicken stock
premium stock
butter
red and green bell pepper (diced)
breadcrumbs
salt

MINCED SHRIMP FILLING:

600 g medium marine shrimps
20 g fatty pork (diced)
1 egg white
salt
sugar
caltrop starch

METHOD:

MINCED SHRIMP FILLING:

1. Shell and devein the shrimps. Rub them in coarse salt. Rinse well. Wipe dry with a clean dry towel.
2. Put the shrimps on a chopping board. Crush the shrimps with the flat side of a cleaver. Coarsely chop them with the back of knife.
3. Put the minced shrimps into a mixing bowl. Add egg white and stir in one direction with a bare hand until sticky. Add salt, sugar, caltrop starch and diced fatty pork. Mix well. Lift the minced shrimp mixture off the bowl and slab it back into the bowl repeatedly until sticky and resilient. This is the filling.

ASSEMBLY:

1. Pick the crab roe from the mud crabs. Cook it in chicken stock. Drain and wait till it cools off. Roll the crab roe into a ball.
2. Wrap the crab roe in minced shrimp filling. Roll it into a ball. Coat in breadcrumbs. Deep-fry in hot oil until golden and crispy. Drain.
3. Remove the head of the lobster. Steam it until cooked through. Save it as garnish.
4. Shell the lobster tail. Cut into chunks. Blanch in warm oil until medium done. Drain.
5. Heat premium stock and butter in a wok. Put in green and red bell pepper. Toss well. Add the blanched lobster chunks. Toss again. Season with salt. Stir in caltrop starch slurry. Save the mixture on a serving plate.
6. Arrange the deep-fried shrimp ball next to the sautéed lobster. Garnish with the steamed lobster head from step 3. Serve.

MASTER FU'S TIPS:

- To make minced shrimp filling, it's best to use male shrimps. Their flesh tends to be more bouncy.
- When you blanch the lobster chunks in warm oil, do not overcook them. They only need to be medium-done at this point because they will be cooked again with other condiments later on. Otherwise, they will be dry and rubbery at last.

Pan-fried M7 Wagyu

*Chinese version refer to p.106

INGREDIENTS:

200 g boneless wagyu ribs
20 g demi-glace
salad greens

MARINADE: (FOR 6 KG OF WAGYU RIBS)

mirepoix stock
salt
sugar
6 eggs
toasted peppercorns

METHOD:

1. Add the marinade to the wagyu ribs. Mix well and leave them overnight.
2. Heat a grilling pan. Put in the wagyu ribs and sear until medium-well done. Slice and save on a serving plate.
3. Heat the demi-glace. Drizzle over the sliced wagyu ribs. Serve with salad greens on the side.

MASTER FU'S TIPS:

- Make sure you don't overcook the beef – it only needs to be medium-well done. It should still have pink centre, with juicy, silky meat.
- To make mirepoix stock, just cut onion, carrot and celery into chunks. Stir-fry in oil until fragrant. Add bay leaves and star-anise. Toss again. Drizzle with wine and add water. Boil for 2 hours.

Baked Stuffed Conch

*Chinese version refer to p.108/makes 1 serving

INGREDIENTS:

1 conch (240 g)
diced cooked chicken
diced grilled chicken liver
diced button mushrooms
diced straw mushrooms
butter
whipping cream
1 tsp curry powder
40 g premium stock
1 egg yolk
1 conch shell

AROMATICS:

grated garlic
chopped shallot
finely diced ginger
diced onion

METHOD:

1. Put the conch in the boiling water until done. Pick out the meat from the shell and finely diced.
2. Heat oil in a wok. Stir-fry aromatics and curry powder until fragrant. Add premium stock and bring to the boil. Add diced conch, diced cooked chicken, diced chicken liver, diced button and straw mushrooms. Toss until fragrant. Add butter and whipping cream. Stir and cook until the mixture reduces slightly. This is the filling.
3. Stuff the conch shell with diced conch meat first. Then top with the filling from step 2.
4. Brush whisked egg yolk over the filling. Bake in a preheated oven at 220°C for 12 minutes until golden. Serve.

MASTER FU'S TIPS:

Conch meat is springy in texture and loaded with umami. It is the perfect match with chicken, chicken liver and mushrooms. You get to taste mouthful of conch meat while savouring the complex textures and layers of flavours.

Stewed Lamb in Clay Pot with Bamboo Shoot and Shiitake Mushrooms

*Chinese version refer to p.110

INGREDIENTS:

1.8 kg lamb brisket
100 g dried shiitake mushrooms
240 g winter bamboo shoot
240 g deep-fried tofu skin sticks
2 whole dried tangerine peels (finely shredded)
ginger
spring onion
100 g garlic cloves
2 tbsp lamb stew sauce
Shaoxing wine

METHOD:

1. Soak dried shiitake mushrooms in water until soft. Cut off the stems. Peel the outer leaves off the winter bamboo shoot. Rinse and cut into chunks. Soak dried tangerine peels in water until soft. Scrape off the pith.
2. Rinse the lamb brisket. Blanch in boiling water in whole. Drain. Brush dark soy sauce all over.
3. Heat oil in a wok. Deep-fry the lamb brisket until golden. Drain. Then shallow-fry in oil again to remove the gamey taste. Rinse well.
4. Heat oil in a wok. Stir-fry ginger, spring onion and garlic cloves until fragrant. Put in lamb brisket, shiitake mushrooms and bamboo shoot. Toss well and drizzle with wine. Stir in the lamb stew sauce. Mix well and add hot water. Bring to the boil and turn to low heat. Simmer for 90 minutes until the lamb is tender.
5. Cook deep-fried tofu skin sticks with sauce until soft. Put them on the bottom of a clay pot. Slice the lamb brisket and arrange over the tofu skin sticks in the clay pot. Put some sauce from the stewed lamb from step 4 into a small pot. Heat it up and stir in caltrop starch slurry. Cook until it thickens. Drizzle over the lamb in the clay pot. Heat up the whole pot and serve.

MASTER FU'S TIPS:

- Lamb brisket needs to be cooked for a long time. It's advisable to line the bottom of your pot with a bamboo mat before stir-frying the aromatics and veggies. That would prevent the lamb from sticking to the pot and burning. You should still make sure there is enough sauce in the pot throughout the cooking process.
- Stew the lamb brisket in whole to retain the juices. Slice it at last and drizzle with a thickened glaze. This is a dish bursting with lamb flavour.

Peony Giant Shrimp

*Chinese version refer to p.112/makes 1 serving

INGREDIENTS:

1 large shrimp (about 100 to 150 g)
1 slice Yunnan ham (finely shredded)

MARINADE:

edible lye
baking soda

EGG WHITE CUSTARD:

1 part egg white
1 part milk
1 part premium stock

METHOD:

1. Rinse, shell and devein the shrimp. Cut off the surface layer of the flesh with a knife. Then make cuts along the length to slice it thinly without cutting all the way through. The shrimp flesh should open up like a book along the back.
2. Mix edible lye and baking soda. Add to the shrimp and leave it for 2 hours. Rinse until a running tap until the shrimp no longer feels slippery.
3. Blanch the shrimp in boiling water. Drain. Blanch the shrimp in hot oil briefly until the shrimp turns snow white.
4. Mix the egg white custard ingredients together. Strain it and pour into a steaming dish. Steam until set.
5. Arrange the shrimp on top. Drizzle with thickened premium stock. Sprinkle with shredded Yunnan ham. Serve.

MASTER FU'S TIPS:

- Peony shrimp is supposed to be snow white in colour. Thus, it's essential to cut off the surface layer of the shrimp that contains the red pigment known as astaxanthin.
- Marinate the shrimp with edible lye and baking soda, to make the shrimp flesh bouncy in texture.

Baked Oyster with Supreme Soy Sauce

*Chinese version refer to p.114

INGREDIENTS:

8 large oysters (shucked)
15 g supreme soy sauce
spring onion (cut into short lengths)
Shaoxing wine

EGG WHITE BATTER:

2 egg whites
corn starch
water

METHOD:

1. Rub oysters in caltrop starch. Rinse well. Soak them in boiling water for 4 minutes. Drain.
2. Dip the oysters in the egg white batter. Deep-fry in hot oil until golden. Set aside.
3. Heat oil in a wok. Stir-fry spring onion until fragrant. Put in the fried oysters and drizzle with supreme soy sauce. Toss well and sprinkle with Shaoxing wine. Serve.

MASTER FU'S TIPS:

You only need to coat the oysters thinly in the egg white batter. You don't want them in a thick crust.

Steamed Iberico Pork Patty with Squid

*Chinese version refer to p.116

INGREDIENTS:

200 g premium pork shoulder butt
25 g dried squid
25 g diced shiitake mushrooms

MARINADE:

1 egg white
2 g salt
2 g sugar
4 g caltrop starch
light soy sauce
dark soy sauce
cooking oil
water

METHOD:

1. Rinse the pork and dice it. Then chop finely with a knife.
2. Peel off the skin on the dried squid. Soak it in water until soft. Finely dice it.
3. Put chopped pork, diced dried squid, diced shiitake mushrooms and marinade into a mixing bowl. Stir clockwise until thick and sticky. Transfer onto a steaming dish.
4. Boil water in a steamer or wok. Steam the pork patty over high heat for 6 minutes until cooked through. Serve.

MASTER FU'S TIPS:

There is only 450 g of premium pork shoulder butt in each pig. It's a rare cut. Master Fu doesn't use a food processor, but chooses to chop the pork by hand. The pork patty will end up tasting richer and juicier; it will be more bouncy in texture. He puts a new spin on the traditional recipe.

Deep-fried Seafood Gold Coins

*Chinese version refer to p.118

INGREDIENTS:

3 to 4 slices sandwich bread
mashed potato
12 to 16 thin slices of shiitake mushrooms
3 to 4 coriander leaves
grated Jinhua ham
1 egg white

FILLING:

28 g diced lobster flesh
28 g diced water chestnut
28 g diced chives

SEASONING:

salt
caltrop starch
egg white

METHOD:

1. Cut each slice of sandwich bread into a disc using a cookie cutter. Mix the filling ingredients together.
2. Add seasoning to the mashed potato. Mix well and stir until smooth.
3. Put 20 g of filling on a disc of bread. Spread mashed potato on top. Sprinkle with shiitake mushrooms, coriander leaves and Jinhua ham.
4. Brush whisked egg white over the stuffed disc of bread. Deep-fry in hot oil until golden. Serve.

MASTER FU'S TIPS:

- You may use any seafood or veggies for the filling - crabmeat, shrimp, yam bean or even yellow chives. You can customise it to your liking.
- After putting the filling on the disc of bread, seal it with a layer of mashed potato. That would seal in the juices and seafood flavour. The mashed potato also prevents the filling from leaking when deep-fried.

Live Shrimp with Supreme Soy Sauce

*Chinese version refer to p.122

INGREDIENTS:

600 g large shrimps (about 75 to 100 g each)
supreme soy sauce
finely diced ginger
chopped shallot
grated garlic
finely chopped spring onion

METHOD:

1. Make a cut along the back of each shrimp without cutting all the way through. Devein, but keep the shell and head on. Rinse well.
2. Heat oil in a wok. Stir-fry ginger, shallot, garlic and spring onion until fragrant. Put in the shrimps. Drizzle with supreme soy sauce. Toss until the shrimps are medium-well done. Mix well and serve.

MASTER FU'S TIPS:

- Master Fu makes a cut along the back of each shrimp so that the shrimps can pick up the flavours of the aromatics and the soy sauce more easily.
- Alternatively, you may remove the shrimp heads and deep-fry them separately. When the shrimp tails are medium-well done, put the shrimp heads back in before drizzling with supreme soy sauce.

Braised Mixed Vegetables with Bamboo Piths

*Chinese version refer to p.124

INGREDIENTS:

300 g elm ear fungus, yellow ear fungus, shiitake mushrooms, straw mushrooms, button mushrooms, winter bamboo shoot, bamboo piths, and hairy moss
mung bean sprouts (with both ends cut off)
4 leafy green stems
premium stock

METHOD:

1. Soak elm ear, yellow ear, shiitake, and bamboo piths in water.
2. Soak hairy moss in water for 20 minutes. Stir in some oil and rub it gently. Rinse with water three to four times.
3. Blanch elm ear, yellow ear, shiitake, straw mushrooms, button mushrooms, winter bamboo shoot and bamboo piths in boiling water twice. Drain.
4. Heat oil in a wok. Stir-fry all mushrooms, bamboo piths and bamboo shoot until fragrant. Cook them in premium stock until flavourful. Drain and set aside.
5. Blanch leafy green stems and mung bean sprouts in boiling water. Drain.
6. Cook all mushrooms, hairy moss, bamboo shoot and bamboo piths in premium stock until flavourful. Arrange on serving plate. Arrange leafy green stems and mung bean sprouts around the plate. Serve.

MASTER FU'S TIPS:

- Elm ear and yellow ear fungus tend to be harder in texture. They have to be cooked in premium stock until soft and flavourful.
- Vegan readers may use vegetarian stock instead of premium stock.

Braised Beef Brisket with Turnip

*Chinese version refer to p.126

INGREDIENTS:

1 slab beef brisket (about 1.2 kg)
600 g beef tendon
white turnip
3 cloves star-anise
sliced ginger
spring onion (cut into lengths)
premium stock

SAUCE:

40 g bean stew sauce for lamb and duck
40 g Chu Hou sauce
10 g fermented soybean sauce
15 g rock sugar

METHOD:

1. Rinse the beef tendon. Blanch in boiling water. Clean well. Steam for 75 minutes until soft.
2. Rinse the beef brisket. Blanch in boiling water. Sear in a little oil until browned on all sides. Set aside.
3. Heat oil in a wok. Stir-fry ginger and spring onion until fragrant. Add bean stew sauce for lamb and duck, Chu Hou sauce and fermented soybean sauce. Toss until fragrant. Put in the beef brisket and tendon, premium stock and rock sugar. Bring to the boil and turn to low heat. Simmer for 75 minutes.
4. Cook chunks of turnip in second stock for 15 minutes until flavourful. Transfer into a clay pot. Slice the beef brisket and tendon. Arrange over the turnip in a clay pot. Pour some braising stock into a pot and stir in caltrop starch slurry. Pour over the beef and turnip in the clay pot. Heat up the whole clay pot and serve.

MASTER FU'S TIPS:

- Bean stew sauce for lamb and duck is actually soybean paste with ginger, garlic and spices. It works well with the rich soy flavour of Chu Hou sauce, and the fermented depth of soybean sauce. They round out each other's flavours and add depth to the beef brisket.
- The sauces should be cooked with fried aromatics in order to accentuate their aromas.

Fruit Salad with Shrimps

*Chinese version refer to p.128

INGREDIENTS:

900 g medium shrimps
100 g iceberg lettuce
30 g celery
2 apples
1 pineapple
2 mangoes
6 strawberries
6 cherries

SALAD DRESSING:

240 g miracle whip
2 hard-boiled egg yolks
1 lemon (squeezed)
salt
condensed milk

METHOD:

1. Mix all salad dressing ingredients together.
2. Peel all fruits; core and dice them. Tear off the tough veins from the celery. Rinse the lettuce and cut into chunks.
3. Blanch the shrimps in boiling water until done. Shell, devein and cut them into halves along the length.
4. Arrange the fruits in a salad dish. Drizzle with the salad dressing and mix well. Put cooked shrimps on top. Serve.

MASTER FU'S TIPS:

- This salad dressing is richly flavoured - with eggy aromas, citric tartness, and sweetness from the condensed milk. It works well with the fruity assortment.
- Fruits and shrimps are arranged in layers. Their colours are highly appetising.

Master Fu's Signature Fried Rice

*Chinese version refer to p.132

INGREDIENTS:

140 g shiba shrimps
20 g finely shredded Jinhua ham
2 eggs (whisked)
76 g spring onion (finely chopped)

STEAMED RICE:

200 g rice
200 g premium stock
cooking oil

METHOD:

1. To make steamed rice, mix rice, cooking oil and premium stock together. Transfer into a steaming bowl. Steam over boiling water until cooked. Fluff the rice while still hot. Leave it to cook. For this recipe, you'd need 320 g of steamed rice.
2. Shell the shrimps. Rinse and blanch in boiling water. Drain. Stir-fry in oil briefly. Set aside. Put finely chopped spring onion on a paper towel. Fold the paper towel over the spring onion to absorb excessive moisture.
3. Heat oil in a wok. Pour in the whisked eggs and toss until half set. Put in the steamed rice and press it flat with a spatula so that all rice grains are heated evenly. Sprinkle with finely shredded Jinhua ham. Toss over high heat until the rice is cooked through. Add shrimps. Sprinkle with spring onion. Toss to mix well and heat through. Serve.

MASTER FU'S TIPS:

- Master Fu heats up the wok till red hot when making this dish. His fried rice isn't greasy, but each grain is separable from others, boasting exceptional wok hei, and a faint fragrance of the premium stock. Make sure you keep tossing the ingredients in the wok continuously. Otherwise, the rice may burn.

- For this recipe, steam the rice in premium stock shortly before you fry the rice. Do not use day-old rice as it's too stiff for this recipe.
- Master Fu prefers shiba shrimps because their texture is more velvety than other shrimps.

Fried Noodles with Shredded Beef Tenderloin

*Chinese version refer to p.134

INGREDIENTS:

120 g beef tenderloin (shredded)

100 g mung bean sprouts (with both ends cut off)

100 g salted mustard green, green and red bell pepper

1 nest Chow Mein noodles

SEASONING:

300 g premium stock

dark soy sauce

salt

caltrop starch slurry

METHOD:

1. Shred the salted mustard greens, red and green bell pepper.
2. Cook the noodles in a pot of boiling water. Stir with chopsticks until the noodles are no longer clumped together. Rinse under a running tap to remove the edible lye. Drain.
3. Heat oil in a wok. Put in the noodles and fry until both sides lightly browned and crispy. Cut into 6 to 8 pieces. Arrange on a plate.
4. Heat oil in a wok. Stir-fry beef tenderloin, salted mustard greens and bell peppers. Add premium stock, dark soy sauce and salt. Stir in caltrop starch slurry and cook until it thickens. Pour the mixture over the fried noodles on a serving plate. Serve.

MASTER FU'S TIPS:

- For the best result, make sure you fry the noodles until crispy. You need to fry them in the right amount of oil and press them with a spatula in the frying process.
- Chow Mein noodle is available in the shop selling noodle, weighing 113-150 g for each.

Fried Rice Vermicelli with Crabmeat

*Chinese version refer to p.136

INGREDIENTS:

320 g rice vermicelli
80 g cooked crabmeat
140 g mung bean sprouts (with both ends cut off)
40 g shredded onion
2 eggs
premium stock
30 g spring onion (finely chopped)

SEASONING:

dark soy sauce
salt

METHOD:

1. Put the rice vermicelli in a pot of cold water. Put it on low heat. Bring to the boil slowly. Stir with chopsticks to separate the rice vermicelli. Drain and drizzle with a few drop of oil. Toss well. Wrap them in towel.
2. Stir-fry shredded onion until done. Put in mung bean sprouts and premium stock until medium done. Drain and set aside.
3. Heat oil in a wok. Pour in whisked eggs and toss until medium-well done. Add rice vermicelli. Fry until fragrant. Sprinkle with salt and dark soy sauce. Toss to mix well. Add cooked crabmeat, shredded onion and mung bean sprouts. Sprinkle with finely chopped spring onion. Toss again and serve.

MASTER FU'S TIPS:

- Master Fu uses rice vermicelli from Heyuan, Guangdong, because they are crisper and silkier in texture. They also hold their shape well throughout the cooking process.
- After blanching the rice vermicelli, toss them in oil and wrap them in a towel. The residual heat will soften the rice vermicelli that way.
- Before saving it on a serving plate, toss rice vermicelli with a spatula dipped in oil for 5 seconds. That would enhance the aromas of the dish.

Rice Noodles with Wagyu Beef and Black Truffle

*Chinese version refer to p.138

INGREDIENTS:

160 g Australian wagyu chuck steak
380 g ribbon rice noodles
25 g green and red bell pepper (shredded)
38 g mung bean sprouts (with both ends cut off)
spring onion (shredded)
shaved black truffle

SEASONING:

light soy sauce
dark soy sauce

METHOD:

1. Remove the tough tendons on the chuck steak. Slice thinly. Arrange neatly on the bottom of a small bowl.
2. Heat oil in a wok. Stir-fry mung bean sprouts until just cooked. Set aside.
3. In the same wok, stir-fry shredded bell peppers and ribbon rice noodles until lightly browned. Drizzle with seasoning. Toss well. Add mung bean sprouts. Toss again. Transfer the mixture into the small bowl with sliced wagyu.
4. Turn the small bowl upside-down on a large plate. Remove the small bowl. Use a kitchen torch to burn the sliced wagyu until medium-well done. Garnish with shaved black truffle and spring onion. Serve.

MASTER FU'S TIPS:

- Just burn the wagyu with a kitchen torch until medium-well done. You can keep it velvety and soft that way.
- Stir-fry the mung bean sprouts until just cooked – they should still be firm, without turning watery or being overcooked.

Fried Rice with Fresh Crabmeat and Conpoy, Wrapped in Lotus Leaf

*Chinese version refer to p.140

INGREDIENTS:

28 g freshly picked crabmeat
120 g shiba shrimps
50 g conpoy
40 g char siu pork (diced)
40 g roast duck meat (diced)
60 g diced bamboo shoot
38 g diced shiitake mushrooms
1 lotus leaf
380 g steamed rice
Shaoxing wine
chicken fat

STEAMED RICE:

200 g rice
200 g premium stock
cooking oil

METHOD:

1. Mix all steamed rice ingredients. Transfer into a steaming bowl. Steam until the rice is cooked through. You'll need 320 g of steamed rice for this recipe.
2. Shell the shrimps. Rinse well. Soak conpoy in water until soft. Drain and tear into fine shreds.
3. Rinse the lotus leaf. Blanch in boiling water. Drain.
4. Soak crabmeat, shrimps, conpoy, shiitake, and bamboo shoot in hot premium stock until flavourful.
5. Heat oil in a wok. Stir-fry char siu pork, roast duck meat, and the crabmeat mixture from step 4 until fragrant. Add rice and toss well. Drizzle with Shaoxing wine and chicken fat.
6. Lay flat the lotus leaf on a counter. Put a few piece of crabmeat and some conpoy on the lotus leaf. Then top with the rice mixture from step 5. Wrap it up and steam for 20 minutes. Turn the rice wrapped in lotus leaf out onto a dish with the seam side down. Cut open the lotus leaf to show the crabmeat and conpoy. Serve.

MASTER FU'S TIPS:

- To make the steamed rice, the amount of rice to premium stock should be one to one by weight. If you use 600 g of rice, use 600 g of premium stock.
- Blanch the lotus leaf in hot water to soften it. It also helps accentuate its aromas.

Pan-fried Turnip Cake

*Chinese version refer to p.142

makes 600 g of turnip cake

INGREDIENTS:

4.2 kg white turnip (before peeling)
120 g dried shrimps
8 Cantonese preserved pork sausage
4 Cantonese preserved liver sausage
conpoy
toasted white sesames
1.6 kg premium stock

BATTER MIX:

600 g long-grain rice flour
160 g corn starch
160 g wheat starch
800 g premium stock

SEASONING:

35 g raw cane sugar slab
40 g salt

METHOD:

1. Rinse the turnip and peel it. Then shred it coarsely.
2. Dice the pork sausage and liver sausage. Finely chop the dried shrimps. Soak conpoy in water until soft. Tear into fine shreds.
3. Make the batter by mixing all ingredients until lump-free.
4. Heat oil in a wok. Stir-fry pork sausage, liver sausage and dried shrimps until fragrant. Add shredded turnip, raw cane sugar slab, salt and the remaining premium stock. Bring to the boil. Pour the mixture into the batter from step 3. Mix well.
5. Transfer the batter into a steaming pan. Steam for 1 hour over high heat. Sprinkle with conpoy and sesame on top. Steam for a few more minutes. Remove from heat and leave it to cool completely. Slice and fry in some oil until golden. Serve.

MASTER FU'S TIPS:

- For this recipe, Master Fu makes a premium stock with old chicken, Jinhua ham and lean pork. Then he uses it to make the batter instead of water. He also cook the shredded turnip in the same stock to infuse it with more flavour.
- Choose the thick turnip that are moist inside. You can taste the sweet and delicious turnip cake.
- The turnip should be shredded with a knife by hand, instead of using a mandolin slicer. The turnip will be cut into different thicknesses naturally that way. It can be cooked through easily, while still retain the crunchy texture of the thicker pieces.

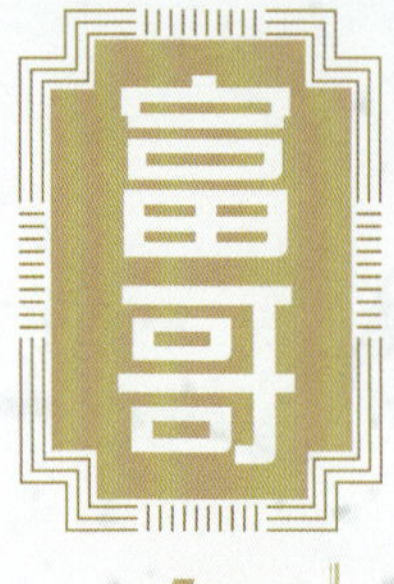

旨人坊

細嘗粵菜真味

著者
鄭錦富 · 方曉嵐

項目統籌
方曉嵐

責任編輯
簡詠怡

攝影
梁細權

裝幀設計
羅美齡

排版
羅美齡 · 楊詠雯

出版者
萬里機構出版有限公司
香港北角英皇道 499 號北角工業大廈 20 樓
電話：2564 7511　　傳真：2565 5539
電郵：info@wanlibk.com
網址：http://www.wanlibk.com
　　　http://www.facebook.com/wanlibk

發行者
香港聯合書刊物流有限公司
香港荃灣德士古道 220-248 號荃灣工業中心 16 樓
電話：2150 2100　　傳真：2407 3062
電郵：info@suplogistics.com.hk
網址：http://www.suplogistics.com.hk

承印者
中華商務彩色印刷有限公司
香港新界大埔汀麗路 36 號

出版日期
二〇二五年七月第一次印刷

規格
大 16 開（190 mm × 253 mm）

ISBN 978-962-14-7620-3

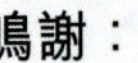

鳴謝：
· 希信（亞洲）有限公司
（瓷器贊助）
· 何宗昇